REGULARIZED RADIAL BASIS FUNCTION NETWORKS

REGULARIZED RADIAL BASIS FUNCTION NETWORKS

Theory and Applications

Paul V. Yee
PMC-Sierra, Inc.

Simon Haykin
McMaster University

A WILEY-INTERSCIENCE PUBLICATION

JOHN WILEY & SONS, INC.

New York / Chichester / Weinheim / Brisbane / Singapore / Toronto

This book is printed on acid-free paper. ♾

Published simultaneously in Canada.

Library of Congress Cataloging-in-Publication Data is available.

Yee, Paul.
Regularized radial basis function networks : theory and applications / Paul Yee, Simon Haykin.
p. cm. - - (Adaptive and learning systems for signal processing, communications, and control)
ISBN 0-471-35349-3 (cloth : alk. paper)
1. Neural networks (Computer science) 2. Feedforward control systems. I. Haykin, Simon S., 1931 - II. Title. III. Series.

QA76.87.V36 2001
006.3′2--dc 21 00-043579

10 9 8 7 6 5 4 3 2 1

To my family
and
To all dedicated scholars

CONTENTS

PREFACE

This book is intended to serve as a bridge between the two areas of nonparametric estimation and artificial neural networks (ANNs). The growing importance and richness of ideas in both areas cannot be ignored and, to that end, this book examines their interplay under the overarching principle of regularization. The specific vehicle for this study is the regularized *strict interpolation radial basis function (RBFN)* estimate or neural network. Aside from its practical importance as one of the better known kernel-based methods for estimation and function approximation, the regularized strict interpolation RBFN is chosen for its straightforward structure, which is simple enough to admit theoretical analysis yet sufficiently powerful for nontrivial applications.

The organization of this book is therefore guided by these two facets. Chapter 1 provides a theoretical understanding of the strict interpolation RBFN, specifically with regards to how its mean-square (MS) consistency can be related to that of the Nadaraya–Watson regression estimate (NWRE), a fundamental kernel-based nonparametric estimate. Regularization in the form of *cross-validation* and *asymptotically optimal* regularization parameter sequences plays a pivotal role in linking these two estimates. Once this background is set, only minor modifications and extensions are required to explain the application of the regularized strict interpolation RBFN to a variety of challenging tasks. As illustrative examples of such tasks, the book discusses probability estimation (and, hence, pattern classification), nonlinear time-series prediction and state estimation, and the dynamic reconstruction of chaotic processes. Chapter 2 indicates briefly how the MS consistency results of Chapter 1 can be used to prove the Bayes risk consistency of the approximate Bayes decision rules formed from regularized strict interpolation RBFN posterior probability estimates. A similar extension is made in Chapter 3 to establish the MS consistency of regularized strict interpolation RBFNs for nonlinear autoregressive time-series prediction, along with corresponding experimental results on speech prediction. Nonlinear time series continue to be the theme for Chapters 4 and 5, where regularized strict interpolation RBFNs are applied to the problems of nonlinear state estimation and dynamic reconstruction of chaotic processes, respectively. Overall, it is hoped that this selection of topics can give the reader an appreciation for the capabilities of the regularized strict interpolation RBFN in a broad range of real-world applications.

The choice to present theory before practice reflects undoubtedly my bias towards a unified logical development of the material. Although I believe that it is easiest to understand the book's material when presented in this order, I also recognize and encourage those readers with more practical inclinations to browse the book, beginning with the Introduction and skipping between the desired application areas in Chapters 2 to 5. In this mode, the theoretically oriented Chapter 1 can act as a reference when the reader desires a more thorough understanding. To assist the reader's mental navigation in either mode, this book introduces specialized or otherwise less well-known concepts on a "just-in-time" basis with brief text digressions and footnotes.

The target audience for this book comprises researchers, practitioners, and graduate students in engineering and the sciences who are interested in nonparametric estimation and ANNs. Given the objectives of this book and its perspective, it is unavoidable that certain mathematical tools, possibly outside the common realm of graduate engineering disciplines, are necessary for the book's development. Thus, the reader is assumed to have some exposure to elementary measure, integration, and probability theory. Beyond these topics, care has been taken to introduce only the minimum set of such auxiliary mathematical concepts that is required to explain properly the results in this book. For these concepts, some fuller background can be found in the appendices. As is often stated in graduate engineering texts, a degree of "mathematical maturity," that is, the ability to follow mathematical arguments, can allow the reader to grasp the main ideas being discussed despite a lack of formal background.

As with any book, certain topics of interest are omitted. For example, optimizing the regularized strict interpolation RBFN architecture by using only a subset of the available training input data to define the RBFN centers is not discussed in any detail, nor is adaptation of other RBFN parameters such as center location and basis function width during overall network cost function minimization. On the one hand, such omissions acknowledge the availability of existing works on these aspects of RBFN design and, on the other, it could be argued that including these additional details, while important in their own right, does not further the stated purposes of this book. In the end, the bridge between nonparametric estimation and ANNs that this book builds represents but a small step in the path towards a more comprehensive understanding of the regularized strict interpolation RBFN as a principled design choice for RBF neural networks.

Acknowledgements

This work grew out of my doctoral studies completed in 1998 at the Communications Research Laboratory (CRL) at McMaster University, Ontario, Canada, under the able guidance of Dr. Simon Haykin. The speech data for the experiments in Chapter 3 were kindly provided by Dean McArthur. Chapters 4 and 5 arose from research performed jointly with Dr. Eric Derbez and Dr. Sadasivan Puthusserypady, respectively. To them and the many others at the CRL (and elsewhere) with whom I've had the pleasure of collaboration, I owe thanks. The same extends to my colleagues Kevin Hung and Jeremy Benson at PMC-Sierra, Inc., for their help in proofreading the page proofs, and Lisa Van Horn and others at John Wiley and Sons, Inc., for their patient assistance during the production process. Not least of all, I thank my wife Susan for her love and understanding during the writing of this book.

PAUL V. YEE

Vancouver, Canada
January 2001

NOTATIONS

$\mathbb{F}^+$	positive subfield of a field $\mathbb{F}$
$\mathbb{F}^{m\times n}$	class of $m \times n$ matrices with elements in field $\mathbb{F}$
$\boldsymbol{X}, Y, \ldots$	random variables (RVs) denoted by uppercase symbols
$\boldsymbol{x}, y, \ldots$	realizations of RVs denoted by corresponding lowercase symbols
$T_n, \boldsymbol{Z}_n, \ldots$	length n sequences of RVs
$t_n, z_n, \ldots$	realizations of length n sequences of RVs
$\Pr\{A\}$	probability of an event A
$I_A(\bullet)$	indicator function (RV) for event A
$P_{\boldsymbol{X},\boldsymbol{Y}}$	joint probability measure for RVs $\boldsymbol{X}$, $\boldsymbol{Y}$
$\int dP_{\boldsymbol{X},\boldsymbol{Y}}, \int dP(\boldsymbol{x}, \boldsymbol{y})$	integration with respect to joint probability measure for RVs $\boldsymbol{X}$, $\boldsymbol{Y}$
$P_{\boldsymbol{X}\mid\boldsymbol{Y}}$	probability measure for RV $\boldsymbol{X}$ conditioned on RV $\boldsymbol{Y}$
$\int dP_{\boldsymbol{X}\mid\boldsymbol{Y}}, \int dP(\boldsymbol{x}\mid\boldsymbol{y})$	integration with respect to probability measure for RV $\boldsymbol{X}$ conditioned on RV $\boldsymbol{Y}$
$X \sim P$	RV X is distributed according to (measure) P
$\lVert \bullet \rVert_p$	p-norm (Euclidean and $p = 2$ unless otherwise specified)
$[\bullet]_{i=1}^n$	vectorization operator for scalar quantity $\bullet$ indexed by i
$[\bullet]_{i,j=1}^{m,n}$	matrification operator for scalar quantity $\bullet$ indexed by i, j
$\mathrm{diag}[\bullet_i]_{i=1}^n$	$n \times n$ diagonal matrix with ith diagonal element equal to $\bullet_i$

$\text{diag}(\bullet_i, i \in I)$	$\lvert I\rvert \times \lvert I\rvert$ diagonal matrix with ith diagonal element equal to ith element of sequence $\{\bullet_i, i \in I\}$
$[\bullet]_I$	subelement vector of $\bullet$ specified by index set I (scalar if I is a singleton)
$\{\bullet\}_I$	subsequence of $\bullet$ specified by index set I (single element if I is a singleton)
$i : j, i \leqslant j, i, j \in \mathbb{Z}$	index set formed by integers i to j, inclusive
$a : \delta : b, a \leqslant b,$ $0 \leqslant \delta \leqslant b - a, a, \delta, b \in \mathbb{R}$	set of reals from a to b step δ, inclusive
A, A^c	set and its complement
$A \Delta B$	symmetric difference of sets A and B
$a \overset{-}{\underset{+}{\gtrless}} f(b \mp c)$	$f(b - c) \leqslant a \leqslant f(b + c)$, where f is an expression containing $b \mp c$
$\arg\max_{a \in A} f(a)$	argument a that maximizes f over candidate set A
a.o.	asymptotically optimal (asymptotic optimality)
a.e.$(-\mu)$, a.s.$(-\mu)$	almost everywhere, almost surely (with respect to measure μ)
i.i.d.	independent and identically distributed
i.p.$(-\mu)$	in probability (with respect to measure μ)
$a_n \nearrow a$	nondecreasing sequence $\{a_n\}$ with limit a
$a_n \searrow a$	nonincreasing sequence $\{a_n\}$ with limit a
$f(n) = \mathcal{O}(g(n))$	$\exists C > 0$ such that $\lvert f(n)\rvert \leqslant C\lvert g(n)\rvert$ for all n sufficiently large
$f(n) = \Omega(g(n))$	$\exists C > 0$ such that $\lvert f(n)\rvert \geqslant C\lvert g(n)\rvert$ for all n sufficiently large
$\text{supp} f$	support of the function f

INTRODUCTION

In the relatively recent field of *artificial neural networks (ANNs)*, two leading classes of feedforward networks have emerged: the *multilayer perceptron (MLP)*, usually with sigmoidal activation function and trained via the *backpropagation* algorithm, and the *radial basis function network (RBFN)*, often with the ubiquitous *Gaussian kernel*. Although both classes have met with considerable success in applications to such difficult engineering problems as nonlinear process estimation and control, there remains a fundamental gap between the theory and practice of *designing* and *applying* these networks. The former relates to the question of "how ought an RBFN be designed to fulfill a particular task" while the latter concerns itself with "for which tasks are RBFNs a justifiably good choice." Indeed, a review of the literature soon reveals the relative dearth of theory explaining the observed successes of ANNs in these areas. A central goal of this book is to shed light on this situation for the class of RBFNs.

PROBLEM DESCRIPTION

The type of RBFN that we shall be considering is a single-layer, feedforward network $\tilde{f}\colon \mathbb{R}^d \to \mathbb{R}$ that consists of a set of *weights* $\{w_i\}_{i=1}^n$ and a set of *basis functions* $\{G_i\colon \mathbb{R}^d \to \mathbb{R}\}_{i=1}^n$, where n is the number of basis or kernel functions. An important property of the RBFN is that it is a *linearly weighted* network, in the sense that the output is formed as

$$\tilde{f}(\bullet) = \sum_{i=1}^{n} w_i G_i(\bullet) \tag{1}$$

This linear combination of (typically) nonlinear basis functions is key to the RBFN's representational ability while maintaining computational and analytical tractability.

Specifically, the radial nature of the RBFN derives from the choice of basis functions G_i each of whose output is dependent only upon the distance of the input to another predetermined point $\boldsymbol{x}_i \in \mathbb{R}^d$. Such a set of radial basis functions can be derived from a common kernel $G: \mathbb{R}^+ \to \mathbb{R}$ via

$$G_i(\bullet) \triangleq G(\|\bullet - \boldsymbol{x}_i\|) \qquad i = 1, 2, \ldots, n \tag{2}$$

where the $\{\boldsymbol{x}_i\}_{i=1}^n$ are collectively known as the *centers* of the network. The particular norm used can be chosen to reflect prior knowledge regarding the nature of the input space, as we shall see in the following sections.

Henceforth, we focus our attention on the problem of *estimating* an unknown input–output mapping $f: \mathbb{R}^d \to \mathbb{R}$ from a finite set T_N of N examples for that function's behavior. In ANN parlance, this task is commonly referred to as *learning* or *training the network* and T_N is called the *training set*. Three common *a priori* models for T_N are:

1. $T_N \triangleq \{(\boldsymbol{x}_i, y_i = f(\boldsymbol{x}_i))\}_{i=1}^N$ (the *noise-free* case).
2. $T_N \triangleq \{(\boldsymbol{x}_i, y_i = f(\boldsymbol{x}_i) + \epsilon_i)\}_{i=1}^N$, where $\{\epsilon_i\}$ is a partially known random process (the *noisy* case).
3. $T_N \triangleq \{(\boldsymbol{X}_i, Y_i)\}_{i=1}^N$, where the $(\boldsymbol{X}_i, Y_i)$ are random samples with an unknown common stationary joint probability measure P_{XY} and bounded variance (the *stochastic* case). In this case, f is implicitly defined by the task at hand, for example, for minimum mean-square error (MMSE) estimation, f is given by the conditional expectation function $f(\bullet) \triangleq \mathbb{E}[Y|\boldsymbol{X} = \bullet]$.

When necessary to do so explicitly, we shall denote the space of possible realizations of T_N by τ^N. It is clear then that once the RBFN has been selected by some means as an appropriate structure for f, the problem of "learning" is equivalent to one of estimating the "best" set of network parameters, that is, weights, basis functions, centers, norms, and the like, from T_N. Although the meaning of "best" is problem dependent, in the ideal case one would like the optimality to extend to regions in the *graph* of f, that is, the set of ordered pairs $\{(\boldsymbol{x}, f(\boldsymbol{x})): \boldsymbol{x} \in \mathbb{R}^d\}$, for which no examples can be found in T_N (otherwise a simple memorization of T_N would suffice). For example, when the input $\boldsymbol{X}$ is a random variable (RV) described by a distribution P_X and quality is measured with respect to the mean-square (MS) estimation error or *risk* R_2^*, we would like to solve the optimization problem of finding the parameter set $\boldsymbol{\theta}^*$ satisfying

$$\boldsymbol{\theta}^* = \arg\min_{\boldsymbol{\theta} \in \mathbb{R}^m} R_2^*(f, \tilde{f}(\bullet, \boldsymbol{\theta})) \qquad R_2^*(f, \tilde{f}) \stackrel{\Delta}{=} \int (f(\boldsymbol{x}) - \tilde{f}(\boldsymbol{x}))^2 dP_{\boldsymbol{X}}(\boldsymbol{x}) \tag{3}$$

where $\boldsymbol{\theta} \in \mathbb{R}^m$ represents the m free parameters of the network. The ANN term *generalization* refers precisely to the ability of a network to extrapolate its performance from the training set to the entire graph of f (or a significant portion thereof). Note that the ideal generalization measure R_2^* is dependent only on m, assuming a given parameterized function class, target function f, and input distribution $P_{\boldsymbol{X}}$. In practice, however, we shall be estimating $\tilde{f} = \tilde{f}_N$ on the basis of a realized training set t_N of T_N and possibly some *a priori* information, so that the ideal generalization measure R_2^* should be modified to read

$$R_2(f, \tilde{f}_N) \stackrel{\Delta}{=} \int \mathbb{E}_{T_N}[(f(\boldsymbol{x}) - \tilde{f}_N(\boldsymbol{x}, \boldsymbol{\theta}))^2] dP_{\boldsymbol{X}}(\boldsymbol{x}) = \mathbb{E}[(f(\boldsymbol{X}) - \tilde{f}_N(\boldsymbol{X}, \boldsymbol{\theta}))^2] \tag{4}$$

that is, the MS risk averaged over all possible inputs and training sets or *global* MS risk. Of course, other variants of the global MS risk are possible by averaging over only one of the input $\boldsymbol{X}$ or the training set T_N in the expectation. Nonetheless, compared to the ideal generalization measure, this training set-dependent generalization measure may now additionally be a function of N and the joint distribution of T_N (in the noisy and stochastic cases). For any reasonable network training procedure, we would expect R_2 to be nonincreasing with increasing m and N. In practice, $m = m(N)$ (as will be explained in the comments regarding network complexity) so that R_2 is primarily indexed by N, the number of available training data.

Notwithstanding the above modifications, the minimization of even R_2 in Eq. (4) with respect to $\boldsymbol{\theta}$ poses some obvious practical difficulties:

1. The optimal network parameters in $\boldsymbol{\theta}^*$ are dependent upon the behavior of f, the unknown function of interest, outside of t_N, the particular realization of the training set T_N we happen to have available.

2. The joint distribution P_{T_N} of the training data is often not known exactly and must also be inferred from the sample training data in t_N. An estimate of the input (marginal) distribution $P_{\mathbf{X}}$ can then be derived from the estimate of the joint distribution, assuming that stationarity holds.

These two uncertainties imply that in all nontrivial cases $\boldsymbol{\theta}^*$ can only be approximately determined. Of course, the availability of T_N partially offsets this lack of knowledge provided that T_N "adequately" captures the behavior of f and, in the cases of noisy or stochastic T_N, the underlying stochastic processes governing $(\boldsymbol{X}, Y)$. An intuitively obvious solution to this problem is to replace the desired objective or cost function involving f with a *proxy* computable using only a training realization t_N and f. Returning to the above example of the noise-free case, for uniformly weighted (nonprobabilistic) inputs with $dP_{\mathbf{X}} = d\mu$, where μ is the usual Lebesgue measure[1] in $\mathbb{R}^d$, we may replace R_2 with $\tilde{J}_2$, where

$$\tilde{J}_2(\tilde{f}, t_N) \triangleq \frac{1}{N}\sum_{i=1}^{N}(y_i - \tilde{f}(\mathbf{x}_i))^2 \tag{5}$$

As $N \to \infty$, the desire is that the parameter vector $\tilde{\boldsymbol{\theta}}^*$ minimizing the proxy yield a network $\tilde{f}^*$ that converges to the (optimum) network f^* that would have been obtained had the exact cost function been used. In other words, we are interested in the convergence of $\tilde{f}^*$ to f^* in some appropriate mode,[2] for example, pointwise *in probability (i.p.)* or *weakly, almost surely (a.s.)* or *strongly*, and in $L_p(D, P_{\mathbf{X}})$ norm for $1 \le p \le \infty$ and some (compact) set $D \subseteq \mathbb{R}^d$. An estimator that satisfies such a convergence condition in a given mode is said to be *consistent* in that mode. If the consistency holds for all possible distributions $P_{\mathbf{X}}$ (for noisy T_N) or $P_{\mathbf{X}Y}$ (for stochastic T_N), we say that the estimator is *universally* consistent.

We now note an important equivalence between the functional consistency with which we are interested and the more usual notion of *risk consistency* used in the statistics literature. Using our previous notations, risk consistency occurs when the parameter vector $\tilde{\boldsymbol{\theta}}^*$ minimizing the proxy yields a network $\tilde{f}^*$ with *performance* converging to that of the network f^* which is optimal with respect to the exact cost function.

[1]See note 1 in Appendix A.1.
[2]See note 2 in Appendix A.1.

For the case of stochastic T_N and performance measured by the MS output estimation error or risk

$$J_2(f) \triangleq \mathbb{E}[|Y - f(\mathbf{X})|^2] \tag{6}$$

the optimum estimator is well known to be $f^*(\bullet) \triangleq \mathbb{E}[Y|\mathbf{X} = \bullet]$ (assuming $\mathcal{F}$ contains the conditional mean function). Then J_2 *risk consistency* is equivalent to our functional $L_2(\mathbb{R}^d, P_{XY})$ consistency, since

$$\begin{aligned} J_2(\tilde{f}^*) - J_2(f^*) &= \mathbb{E}[\tilde{f}^*(\mathbf{X})^2 - 2\tilde{f}^*(\mathbf{X})Y + Y^2 - f^*(\mathbf{X})^2 + 2f^*(\mathbf{X})Y - Y^2] \\ &= \mathbb{E}[\tilde{f}^*(\mathbf{X})^2] - 2\mathbb{E}[\tilde{f}^*(\mathbf{X})f^*(\mathbf{X})] + \mathbb{E}[f^*(\mathbf{X})^2] \\ &= \mathbb{E}[(f^*(\mathbf{X}) - \tilde{f}^*(\mathbf{X}))^2] \end{aligned} \tag{7}$$

where $(\mathbf{X}, Y)$ are assumed independent of T_N, and we have used the facts that

$$\mathbb{E}[(Y - f^*(\mathbf{X}))f^*(\mathbf{X})] = 0 \qquad \mathbb{E}[g(\mathbf{X})f^*(\mathbf{X})] = \mathbb{E}[g(\mathbf{X})Y] \tag{8}$$

for all Borel measurable functions g[3] such that the expectations exist. Thus for this case of *regression* estimation, there is no loss of generality in studying functional consistency versus output estimation consistency.

Although consistency is an asymptotic property, it is considered one of the most desirable fundamental properties of an estimator and is consequently a key object of study. Note that the functional consistency with which we are concerned is a weaker condition than the consistency of the optimal proxy parameter vector $\tilde{\boldsymbol{\theta}}^*$ with respect to the optimal exact parameter vector $\boldsymbol{\theta}^*$; when f is a continuous function, the latter implies the former but the reverse need not be true.

In summary, given a task and an associated training set T_N, the problem of *design* for a RBFN is one of choosing the network parameters that minimize the risk and thereby maximize the desired generalization performance. Since achieving this aim would require knowledge of the unknown function f being estimated, the next logical goal is to produce estimates $\tilde{f}$ of f using T_N (and possibly some prior knowledge about f) with performance converging in N to that of f as rapidly as possible.

[3]See note 3 in Appendix A.1.

REVIEW OF CURRENT APPROACHES

In the previous section, we described the problem of RBFN design in rather general terms. Here we shall survey the recent history of RBFN design and examine some of the major current approaches found in both the ANN and related statistical literature, particularly from the area of *kernel regression estimates (KRE)* (Eubank, 1988; Härdle, 1990). Again the primary objective is to lay bare the design choices made by each method so that their theoretical properties may be properly studied.

Traditional ANN Approaches

The seminal work of Broomhead and Lowe (1988) is representative of early attempts at RBFN design. Originally inspired by neurobiology, this school typically employs a Gaussian kernel $G(r) = \exp(-r^2)$ and a norm weighting matrix set to $h^{-1}\boldsymbol{I}$, where $h \in \mathbb{R}^+$ is a width parameter that controls the "spread" of each kernel about its center; the suggested empirical formula for h is $h = d^2/n$, where $d = \min_{i,j=1,\ldots,N} \|\boldsymbol{c}_i - \boldsymbol{c}_j\|$ is the minimum Euclidean distance between the $n \leq N$ centers chosen from T_N. This subset selection may be simply random or organized in some other fashion, for example, via clustering. The network weights $\tilde{\boldsymbol{w}}$ are then computed as the solution to the L_2 interpolation problem

$$\tilde{\boldsymbol{w}} = \arg\min_{\boldsymbol{w} \in \mathbb{R}^n} \|\boldsymbol{y} - \boldsymbol{G}\boldsymbol{w}\|^2 \tag{9}$$

where $\boldsymbol{y} \in \mathbb{R}^N$ is the vector of desired output values in T_N and $\boldsymbol{G} \in \mathbb{R}^{N \times n}$ with entries $[\boldsymbol{G}]_{ij} = G(\|\boldsymbol{x}_i - \boldsymbol{c}_j\|/h)$ is the *interpolation matrix* for this problem. The required solution for $\tilde{\boldsymbol{w}}$ is the classical pseudo-inverse formula for an overdetermined linear least-squares problem

$$\tilde{\boldsymbol{w}} = \boldsymbol{G}^+\boldsymbol{y} \qquad \boldsymbol{G}^+ \stackrel{\Delta}{=} (\boldsymbol{G}^\top\boldsymbol{G})^{-1}\boldsymbol{G}^\top \tag{10}$$

where $\boldsymbol{G}^+$ denotes the pseudo-inverse of $\boldsymbol{G}$. Somewhat later, motivated by developments in the related area of *splines* for image surface reconstruction in computer vision, Poggio and Girosi (1990) formulated the seminal approach to RBFN design based on the principle of *regularization*. As in splines, the function estimate is derived as the solution to an infinite-dimensional variational problem over a given (Hilbert) space[4] $\mathcal{H}$ of

[4]See note 4 in Appendix A.1.

typically "smooth" functions. The problem is to minimize a linear functional H over $\mathcal{H}$ consisting of two terms: the first term, H_1, being the standard sum-of-squared fitting errors over T_N, while the second term, H_2, is a roughness penalty typically measured through the norm induced by a *pseudo-differential* operator $P: \mathcal{H} \to \mathcal{H}$.[5] More precisely, we set

$$Hf \stackrel{\Delta}{=} H_1 f + \lambda H_2 f \qquad H_1 f \stackrel{\Delta}{=} \sum_{i=1}^{N} (y_i - f(x_i))^2 \qquad H_2 f \stackrel{\Delta}{=} \|Pf\|^2 \tag{11}$$

and seek $\tilde{f}$ satisfying

$$\tilde{f} = \arg\min_{f \in \mathcal{H}} Hf \tag{12}$$

where $\lambda \in \mathbb{R}^+$ is the *regularization* parameter that balances the fidelity of $\tilde{f}$ to T_N with the smoothness of $\tilde{f}$. Setting $\lambda = 0$ yields a standard least-squares solution while $\lambda \to \infty$ yields (as we shall show) an estimate closely related to classical KRE methods. This functional optimization approach has the advantage that with the proper choice of P and $\mathcal{H}$, the unique minimizer of Eq. (11) is precisely the RBFN described by Eq. (1) and can be shown to have the following characteristics (Yee, 1992; Poggio and Girosi, 1990):

1. Each input datum in T_N is a center for the RBFN, that is, $n = N$ and $\mathbf{c}_i = \mathbf{x}_i$, $i = 1, 2, \ldots, n$. This situation is referred to as the *strict interpolation (SI)* case and will be seen to have direct links to the classical KRE methods.
2. For $\mathcal{H} = \mathcal{S}$, the Hilbert space of rapidly decreasing, infinitely continuously differentiable functions found in the Schwartz theory of tempered distributions,[6] and P and its adjoint P^* satisfying

$$P^*P = \sum_{n=0}^{\infty} \frac{(-1)^n}{n! 2^n} \nabla_U^{2n} \qquad \nabla_U^2 \stackrel{\Delta}{=} \sum_{i=1}^{d} \sum_{j=1}^{d} u_{ij} \frac{\partial^2}{\partial x_i \partial x_j} \tag{13}$$

we obtain the important case of the Gaussian kernel with norm-weighting matrix $\boldsymbol{U} \stackrel{\Delta}{=} [u_{ij}]_{i,j=1}^{n}$:

$$G_i(\bullet) = \exp(-\|\bullet - \boldsymbol{x}_i\|_U^2/2) \tag{14}$$

[5] See note 5 in Appendix A.1.
[6] See note 6 in Appendix A.1.

where U is assumed to be symmetric positive definite and $\|\bullet\|_U^2 \triangleq \bullet^\top U \bullet$.

3. The network weights $w \triangleq [w_i]_{i=1}^n$ satisfy the *SI* equation over T_N

$$(G + \lambda I)w = y \tag{15}$$

These optimum network weights can also be obtained by assuming *a priori* that the estimate $\tilde{f}$ lies in the linear span of $\{G_i\}_{i=1}^n$ and seeking the solution to the constrained problem

$$w = \arg\min_{\omega \in \mathbb{R}^n} \|y - G\omega\| \text{ subject to } \|\sqrt{G}\omega\| = c \tag{16}$$

for some $c > 0$ and where $\sqrt{G}$ is a square-root matrix for G. It can be shown that for the Gaussian basis functions G_i in Eq. (14), the interpolation matrix G is always positive definite for any distinct set of centers $\{x_i\}$ so that its square-root exists.[7] By the Lagrange multiplier method, this condition is equivalent to

$$w = \arg\min_{\omega \in \mathbb{R}^n} \left(\|y - G\omega\|^2 + \lambda \|\sqrt{G}\omega\|^2 \right) \tag{17}$$

The network weights have the further property that they are directly proportional to the interpolation errors over T_N through the regularization parameter λ, that is,

$$\lambda w_i = y_i - \tilde{f}(x_i) \tag{18}$$

Although it is clear from this relationship that $\lambda = 0$ yields an exact fit to T_N, the situation in the other limiting case $\lambda \to \infty$ is not as obvious.

Further extensions to the method involve setting $n < N$ and *assuming* Eq. (1) to be the correct form for $\tilde{f}$, that is, $f \in \operatorname{span}(\{G_i\}_{i=1}^n)$. The network weights can then be shown to satisfy

$$(G^\top G + \lambda \Gamma)w = G^\top y \tag{19}$$

where $G = [G_j(x_i)]_{i,j=1}^{N,n}$, $\Gamma = [G_j(c_i)]_{i,j=1}^n$, and $G_i(\bullet) \triangleq G(\|\bullet - c_i\|_U)$, $i = 1, 2, \ldots, n$. Note these definitions are general in that we obtain precisely the pseudo-inverse solution of Broomhead and Lowe (1988)

[7]See note 7 in Appendix A.1.

by setting $\lambda = 0$ and the SI case by letting $\boldsymbol{c}_i = \boldsymbol{x}_i$, $i = 1, 2, \ldots, n = N$. We may allow even greater network flexibility by optimizing Eq. (11) simultaneously with respect to all free network parameters, that is, weights, centers, and the entries of the norm-weighting matrix, leading to what Poggio and Girosi (1990) call the *hyperbasis function (HyperBF)* method. The original work proposes using simple gradient descent to perform the minimization, although more sophisticated optimization methods such as conjugate-gradient search (Hutchinson, 1994) have also been used more recently.

Limitations of the ANN Approaches

Since the approach of Poggio and Girosi (1990) subsumes that of Broomhead and Lowe (1988) as a special case, we shall concentrate on the former. In the simplest case of $\lambda = 0$, we see that $H_1 f = N\tilde{J}_2(f, T_N)$, the sum-of-squared fitting errors over T_N, is being used as a (scaled) proxy for the desired generalization measure R_2 described previously. Justification for such proxies in the general case can be found in the principle of *empirical risk minimization (ERM)*. Basically, ERM follows the intuitive idea of minimizing a proxy for the risk based on empirical data when the actual risk cannot be computed due to incomplete knowledge. This approach is supported by various theories, of which we shall consider below the theory of *uniform convergence of empirical averages to their mathematical expectations* as developed by Vapnik (1982). Other related approaches include those using the notion of *uniform strong law of large numbers (USLLN)*, and lead to similar results (Pollard, 1984; Gallant, 1987).

For noise-free T_N consisting of independent, identically distributed (i.i.d.) examples $y_i = f(\boldsymbol{x}_i) \in \{0, 1\}$, this uniform convergence is characterized by an intrinsic coefficient of the combinatorial complexity of the class of functions $\mathcal{F}$, assumed to contain both f and its approximant $\tilde{f}$.[8] This coefficient is called the *Vapnik–Chervonenkis (VC) dimension* of $\mathcal{F}$ [denoted $d_{\mathrm{VC}}(\mathcal{F})$] (Vapnik, 1982) and can be defined as follows. Consider an arbitrary collection T_N of N examples. Applying a function $f \in \mathcal{F}$ to the example inputs $\{\boldsymbol{x}_i\}$ in T_N results in a *dichotomy* or partition of T_N into those examples whose inputs are mapped by f to 1 and those mapped to 0. We then say that f *induces* a dichotomy over T_N. The complexity of a class

[8]By "approximant," we mean a function that is deemed to approximate another (target) function.

of functions $\mathcal{F}$ can then be measured by the maximum number N for which all possible 2^N dichotomies over an arbitrary example set T_N can be induced by some function in $\mathcal{F}$.

A major result of the theory is that a finite $d_{VC}(\mathcal{F})$ is both necessary and sufficient for a distribution-free, worst-case upper bound on the probability of a large deviation between a given desired generalization measure and its empirical proxy to hold. For the usual case $\mathcal{F} = \mathcal{F}_{\boldsymbol{\theta}}$, a function class parameterized by a vector $\boldsymbol{\theta} \in \mathbb{R}^m$, and letting $\tilde{f}_{\boldsymbol{\theta}}$ denote the function in $\mathcal{F}_{\boldsymbol{\theta}}$ determined by $\boldsymbol{\theta}$, we have the result that for any distribution on $\boldsymbol{X}$,

$$\Pr\left\{ \sup_{\boldsymbol{\theta} \in \mathbb{R}^m} |\tilde{R}(\tilde{f}_{\boldsymbol{\theta}}, T_N) - R(f, \tilde{f}_{\boldsymbol{\theta}})| > \epsilon \right\} < C(d_{VC}, N) \exp(-K\epsilon^2 N) \qquad (20)$$

where $\tilde{R}(\tilde{f}_{\boldsymbol{\theta}}, T_N)$ and $R(f, \tilde{f}_{\boldsymbol{\theta}})$ denote general empirical and actual risk functions, respectively, with $C(d_{VC}, N)$ a function growing at most polynomially in N and K a positive constant (independent of N). Under these conditions, the right-hand side of Eq. (20) vanishes as $N \to \infty$, which means that, with probability approaching one, the empirical risk $\tilde{R}(\tilde{f}_{\boldsymbol{\theta}}, t_N)$ is close to the actual risk $R(f, \tilde{f}_{\boldsymbol{\theta}})$. This "closeness" occurs regardless of the choice of parameter vector $\boldsymbol{\theta} \in \mathbb{R}^m$, that is, *uniformly* over the function class $\mathcal{F}_{\boldsymbol{\theta}}$. When this condition holds, we say that we have the *uniform convergence of empirical averages to their mathematical expectations*, the "empirical average" in this case being the empirical risk and the "mathematical expectation" being the actual risk.

In fact, since the right-hand side of Eq. (20) forms a summable sequence (in N), we can invoke the Borel–Cantelli lemma[9] to conclude that the empirical risk $\tilde{R}(\tilde{f}_{\boldsymbol{\theta}}, t_N)$ converges a.s. to the actual risk $R(f, \tilde{f}_{\boldsymbol{\theta}})$. Consequently, we see that if, for each N, we can find a $\tilde{\boldsymbol{\theta}}_N^*$ that minimizes $\tilde{R}(\tilde{f}_{\boldsymbol{\theta}}, T_N)$, then the sequence of networks $\tilde{f}_{\tilde{\boldsymbol{\theta}}_N^*}$ so obtained has generalization performance converging a.s. to the best possible over all approximating functions, that is, we have a.s. risk consistency.

To justify choosing $\lambda > 0$, Poggio and Girosi (1990) offer a Bayesian interpretation of the cost functional Hf under the scenario of noisy T_N. The first term, $H_1 f$, corresponds to the negative log-likelihood of T_N assuming that the additive measurement noise ϵ_i is white Gaussian, while the second term, $H_2 f$, corresponds to the negative logarithm of the (improper) prior probability $P(f) \triangleq \exp(-\lambda \|Pf\|^2)$, that is, λ determines the "variance" or "spread" of the prior probability over the space of

[9]See note 8 in Appendix A.1.

candidate functions. Solving for the *maximum a posteriori (MAP)* estimate under these assumptions is equivalent to the minimization of Hf. A related point of view can be found in the principle of *minimum description length (MDL)* for model selection (Rissanen, 1978). Roughly speaking, MDL introduces the concept of the "description length" of a model that measures its ability to describe the model data against its inherent complexity. In this view, the description length of a complex model that approximates given data well can be equivalent to that of a simpler model that exhibits greater approximation error over the same data, and in this sense, neither model is particularly preferable to the other. In our case, Hf is the *hypothesis description length/complexity* of the model function f, composed of $H_1 f$, which is the *data description length/complexity*, and $H_2 f$, which is *model description length/complexity*. The role of λ is to balance these two competing quantities based on some *a priori* knowledge or preferences. Both of these interpretations, however, do not offer any explicit procedure to select $\lambda > 0$.

A somewhat more forthright selection criterion for $\lambda > 0$ can be formulated from the principle of *structural risk minimization (SRM)* (Guyon et al., 1992; Vapnik, 1992). Returning to the VC dimension theory for binary-valued functions previously introduced, define the *guaranteed risk* as the sum of the empirical risk $\tilde{R}$ and the deviation ϵ in Eq. (20) expressed as a function of the bounding probability $\alpha \triangleq C(d_{VC}) \exp(-K\epsilon^2 N)$. We can therefore assert

$$\Pr\{R(f, \tilde{f}_{\boldsymbol{\theta}}) < \tilde{R}(\tilde{f}_{\boldsymbol{\theta}}, T_N) + \epsilon(\alpha)\} > 1 - \alpha \tag{21}$$

For given N, the empirical risk is a nonincreasing function of d_{VC} while the deviation function or *confidence interval* $\epsilon(\alpha)$ is nondecreasing in d_{VC}. Thus their sum, the guaranteed risk, achieves its minimum with respect to d_{VC}, say at $d_{VC} = d^*_{VC}$. The principle of SRM is to seek the function $\tilde{f}^*$ with this guaranteed-risk-minimizing VC dimension d^*_{VC} by solving a suitable sequence of minimization problems over a corresponding sequence of subsets of the candidate function space $\mathcal{F}_{\boldsymbol{\theta}}$. The subset sequence is structured to have monotonically increasing VC dimension, hence nonincreasing empirical risk and nondecreasing confidence interval. Within each subset, we find the empirical-risk-minimizing network and select the subset containing the network with smallest overall empirical risk. Over this best subset, we then seek the network with minimum guaranteed risk. The rationale for SRM is that for a fixed quantity of training data T_N, the *capacity* of the network, measured by d_{VC}, should be matched to the amount of information available in T_N to avoid the problems of *underfitting* [in the case $d_{\text{VC}}(\mathcal{F}_{\boldsymbol{\theta}}) < d^*_{\text{VC}}$] and *overfitting* [in the case $d_{\text{VC}}(\mathcal{F}_{\boldsymbol{\theta}}) > d^*_{\text{VC}}$]. In the former case, the candidate function space is underparameterized and therefore lacks sufficient flexibility to model the relationships in T_N, leading to estimator bias. The opposite is true in the latter case where $\mathcal{F}_{\boldsymbol{\theta}}$ contains too many free parameters for the given data in T_N to specify accurately, resulting in excessive estimator variance. For more precise details on these issues, the reader may refer to (Geman et al., 1992).

For the case of linearly weighted networks such as the RBFN, an appropriate structure may be introduced in the form of *weight decay.* Following our previous notation, let $\mathcal{F}_k \triangleq \{\tilde{f}_{\boldsymbol{\theta}} : \|\boldsymbol{\theta}\|_E \leq c_k\}$, $k = 1, 2, \ldots, M$, where $\|\bullet\|_E$ is a suitable weighted Euclidean norm in $\mathbb{R}^m$, and $\{c_k\}_{k=1}^M$ is a strictly increasing sequence of positive reals. In other words, the requirements for SRM can be met with a simple structure that considers increasing subsets of networks with successively "larger" weights (as measured by the sum-of-squared network weights). Note that by the strict subset construction $\mathcal{F}_j \subset \mathcal{F}_k$ for $j < k$, it follows that $d_{\text{VC}}(\mathcal{F}_j) < d_{\text{VC}}(\mathcal{F}_k)$ for $j < k$, that is, the structure's subsets have (strictly) increasing capacity. To minimize the empirical risk over each subset $\mathcal{F}_k$, we may use the Lagrange multiplier technique to seek

$$\boldsymbol{\theta}_k^* \triangleq \min_{\boldsymbol{\theta} \in \mathbb{R}^m} (\tilde{R}(\tilde{f}_{\boldsymbol{\theta}}, T_N) + \lambda_k \|\boldsymbol{\theta}\|_E) \tag{22}$$

and thereby set $\tilde{R}_k^* \triangleq \tilde{R}(\tilde{f}_{\boldsymbol{\theta}_k^*}, T_N)$, where $\{\lambda_k\}_{k=1}^M$ is a strictly decreasing sequence of positive reals. Hence for this structure, the principle of SRM chooses as optimal the regularization parameter $\tilde{\lambda}^* = \lambda_j$, where $\mathcal{F}_j$ satisfies $\tilde{R}_j^* \leq \tilde{R}_i^*$ for $j \neq i$, $i = 1, 2, \ldots, M$. For a sufficiently large number of subsets M, it is hoped that one will contain a network with VC dimension d^*_{VC}.

The basic ERM/SRM framework described above has been significantly extended, particularly in two respects. First, by using the concept of *covering number* (Pollard, 1984) in place of the VC dimension, the range of the unknown function f and its approximating function space $\mathcal{F}$ can be a bounded subset of $\mathbb{R}$ (say A) instead of $\{0, 1\}$. To achieve this generalization, the covering number $\mathcal{N}_\bullet(\epsilon, A)$ of $A \subset \mathbb{R}^d$ is defined as the cardinality of the smallest ϵ *cover* for A, where a set $C \subset \mathbb{R}^n$ is called an ϵ cover of A if each point $a \in A$ is at least ϵ close (in a specified norm $\bullet$) to some point in the ϵ-cover C. Given a set of N input data $\boldsymbol{x}^N \triangleq \{\boldsymbol{x}_i\}_{i=1}^N$, we consider the RV $\mathcal{N}_{\bullet}(\epsilon, \mathcal{F}(\boldsymbol{X}^N))$, where $\mathcal{F}(\boldsymbol{x}^N) \triangleq \{[f(\boldsymbol{x}_i)]_{i=1}^N \in \mathbb{R}^N : f \in \mathcal{F}\}$ is the set of all possible output vectors for functions in the class $\mathcal{F}$ when evaluated at the input points in $\boldsymbol{x}^N$. For example, if we let $\mathcal{F}$ be the class of all $\{0, 1\}$-valued functions and use the ∞ norm, that is, maximum elementwise deviation between two vectors, it can be seen that the VC dimension $d_{\mathrm{VC}}(\mathcal{F})$ is equal to the largest N such that, for all possible input data sets $\boldsymbol{x}^N$ of size N, $\mathcal{N}_{\infty}(\epsilon, \mathcal{F}(\boldsymbol{x}^N)) = 2^N$ for all $\epsilon \in (0, 1/2)$. In the more general case where $\mathcal{F}$ has a bounded output range in $\mathbb{R}^d$, an important quantity analogous to the VC dimension in characterizing the uniform convergence of empirical averages to their mathematical expectations is the *expected value* of $\mathcal{N}_1(\epsilon, \mathcal{F}(\boldsymbol{X}^N))$, where the vector norm used is the average 1-norm over the N-dimensional output vectors in $\mathcal{F}(\boldsymbol{x}^N)$. Specifically, the fundamental form of the inequality Eq. (20) is still valid with $\mathbb{E}[\mathcal{N}_1(\epsilon, \mathcal{F}(\boldsymbol{X}^N))]$ in place of $d_{\mathrm{VC}}(\mathcal{F})$ and with $\mathcal{F} = \mathcal{F}_{\boldsymbol{\theta}}$. While not the only route to such results, the notion of covering number may be considered a first natural extension of the VC theory to continuous-valued functions.

The second extension is similar to the method of SRM. By defining a suitable increasing sequence of candidate function spaces $\mathcal{F}_j \subset \mathcal{F}_k \subset \mathcal{F}$ for $j < k$, we can relax the assumption $f \in \mathcal{F}$ by choosing $\mathcal{F} \triangleq \cup_{k=1}^{\infty} \mathcal{F}_k$ to be dense in a function space large enough to contain f, for example, $L_p(D, P_{\boldsymbol{X}})$ for any distribution $P_{\boldsymbol{X}}$. The general notion of allowing $\mathcal{F}$ to grow with N was studied by Grenander (1981) under the name *method of sieves*[10] while the latter property of density was earlier studied for RBFNs by Park and Sandberg (1991, 1993) and commonly called the *universal approximation property (UAP)* in the ANN literature. The seminal work of Krzyżak et al. (1996) combines these two extensions to provide conditions under which SRM-type RBFNs are J_2 risk consistent, both i.p. and universally a.s. Also given are distribution-free finite sample upper bounds on the probability of a large deviation between the classification error rates of the empirical and actual classification-error-minimizing RBFNs.

Parallel to the general mathematical developments of the ERM/SRM theory, Niyogi and Girosi (1996) directly analyze nonregularized RBFNs trained with ERM and stochastic T_n to prove their consistency by providing probably approximately correct convergence rates on their generalization error $R_2(f, \tilde{f}_{n,m})$, where, as before, m is the number of free network parameters. While sharing the same spirit, the results obtained are not universal in the same way that the ERM results are, for

[10]See note 9 in Appendix A.1.

example, the conditional mean f is assumed to belong to a class of functions generated by the convolution of a Gaussian kernel with a signed Radon measure of bounded total variation.[11] A regularized case is addressed in Corradi and White (1995) where the total squared error of two types of "sufficiently kinky polynomial splines" in one dimension (Stinchombe and White, 1990) is studied.[12] There they show that for noisy T_n with i.i.d. measurement errors and f assumed to lie in a suitable space of smooth functions [specifically a *reproducing kernel Hilbert space (RKHS)*[13] (Aronszajn, 1950; Wahba, 1990)], such regularization networks' error converges at a rate that is optimal with respect to the smoothness of f as measured by the order of its differentiability. The kernel functions considered therein, however, do not include the usual radial cases such as the Gaussian kernel commonly in use.

Despite the theoretical justification for their RBFN design procedures, the ERM-based methods, such as the HyperBF method discussed previously, suffer from a number of limitations:

1. Because they are usually nonlinear in multiple parameters, the empirical risk functions to be minimized are subject to *local minima*. To avoid this difficulty, one may reduce the number of free parameters for the candidate function space $\mathcal{F}$ from that of the most general one considered in the HyperBF method. Even then, unless density arguments are offered, as Krzyżak et al. (1996) do, one cannot be certain that the restricted class is sufficiently broad to include (eventually) the unknown target function, making the argument for consistency tenuous.
2. For the SRM-based methods such as weight decay, it is unclear exactly how the sequences defining the subsets are determined, given a training set T_N. The previous comment on the difficulty of ERM over just a single candidate function space raises questions about the feasibility of repeating the constrained minimizations over each subset as required by SRM.
3. As currently developed, the ERM framework addresses neither the case of *correlated examples* in the training set T_N, as would occur in applications involving *time series*, nor that of *nonstationary distributions* P_{XY}, for example, in the case of noisy T_N with *hetero-*

[11] See note 10 in Appendix A.1.
[12] See note 11 in Appendix A.1.
[13] See note 12 in Appendix A.1.

skedastic errors ϵ_i, that is, the case where the error RVs ϵ_i do not share a common variance.

Given that the ERM approach to the theory of ANN is relatively recent, one can expect these issues to be resolved for RBFNs after further study. Indeed, we would be remiss if we did not mention the work of White (1990) with MLP networks in this regard. There, in an early application of the method of sieves and the concept of *metric entropy*, White (1990) proves the weak MS consistency of single-layer, feedforward MLP networks with complexity determined by the method of *ordinary cross-validation (OCV)*[14] for both i.i.d. and stationary ϕ or α-*mixing* inputs.[15] Since mixing processes are correlated, the issue raised in point 3 has been advanced, and the same is true for the regularization induced by OCV with respect to point 2. Point 1 is answered in part by a result stating the convergence i.p. of network parameters determined by approximate minimization to a neighborhood of the truly optimal network parameters when $N \to \infty$. What we shall show, however, is that from an ANN point of view, there exists another route to justifying the theoretical and practical RBFN design choices by adapting results from classical KRE theory, which is the subject of the next section.

Kernel Regression Approach

With its earlier beginning compared to ANNs, the field of KRE is understandably more mature, especially in terms of theory. KRE has its roots in the ideas of nonparametric *kernel density estimation (KDE)* (Parzen, 1962; Cacoullos, 1966), whose origin can be traced back to the well-known frequentist histogram. The generalization from the histogram's effectively rectangular kernel to other kernel shapes as well as the nonlinear *nearest neighbor* type of density estimators soon followed along with commensurate consistency arguments. Particularly appealing is the simplicity of the KDE, which, given an i.i.d. training set $T_n \stackrel{\Delta}{=} \{\boldsymbol{X}_i\}_{i=1}^n$ with product distribution $P_{T_n} \stackrel{\Delta}{=} \prod_{i=1}^n P_X$, has the general form

$$\tilde{f}_n(\bullet) = \frac{1}{nh_n^d} \sum_{i=1}^{n} K\left(\frac{\bullet - \boldsymbol{X}_i}{h_n}\right) \tag{23}$$

[14]For a definition of the OCV method, see the discussion concerning Eq. (36).
[15]See note 13 in Appendix A.1.

with $K: \mathbb{R}^d \to R$ satisfying

$$0 \le K(\boldsymbol{x}) < \infty \qquad K(\boldsymbol{x}) = K(-\boldsymbol{x}) \qquad \text{for all } \boldsymbol{x} \in \mathbb{R}^d \tag{24}$$

$$\lim_{\|\boldsymbol{x}\| \to \infty} \|\boldsymbol{x}\|^d K(\boldsymbol{x}) = 0 \qquad \int K = 1 \qquad \int \|\boldsymbol{x}\|^2 K(\boldsymbol{x})\, d\mu(\boldsymbol{x}) < \infty \tag{25}$$

where μ indicates Lebesgue measure. By considering the case of the histogram kernel $K(\bullet) \stackrel{\Delta}{=} I_{[-1/2,+1/2]^d}(\bullet)$, it is intuitively clear that $h_n \stackrel{n\to\infty}{\to} 0$ in order for the approximate density to converge (pointwise) to the actual density. On the other hand, since the expected number of training points in T_n falling inside a hypercube with edge h_n centered about a given point $\boldsymbol{x} \in \mathbb{R}^d$ is roughly $nh_n^d f(\boldsymbol{x})$, we should also require that $nh_n^d \stackrel{n\to\infty}{\to} \infty$. Remarkably, it turns out that these two conditions, namely

$$h_n \stackrel{n\to\infty}{\to} 0 \qquad nh_n^d \stackrel{n\to\infty}{\to} \infty \tag{26}$$

are both necessary and sufficient for the L_2 and a.s. pointwise consistency of $\tilde{f}_n$ under some mild assumptions (Györfi et al., 1989; Bosq, 1996).

If K is also radially symmetric so that $K((\bullet - \boldsymbol{X}_i)/h_n) = G_i(\bullet)$ in Eq. (1), we see an immediate similarity between the basic KDE and a restricted form of the strict interpolation RBFN with uniform weights $w_i \stackrel{\Delta}{=} 1/(nh_n^d)$, $i = 1, 2, \ldots, n$, and norm-weighting matrix $\boldsymbol{U}_n \stackrel{\Delta}{=} \boldsymbol{I}/h_n$. Of course, an interpretation of the conditions under which such a solution would reasonably occur is more problematic, as there is usually no set of explicit output targets provided with T_n in density estimation. We shall see later, however, that a RBFN trained using the class indicator values as the output targets in T_n corresponds to L_2 estimates of posterior probabilities. We should further mention that the KDE can also be derived as the solution to an appropriately regularized functional minimization problem similar to that for the strict interpolation RBFN. The description here follows that of Section 9.9 in Vapnik (1982). Restricting to the case of estimating a one-dimensional density $f \in L_2([a, b])$ over some (nonempty) interval $[a, b] \subset \mathbb{R}$ and recognizing that the cumulative distribution function F for f satisfies the functional equation

$$F(x) = (Lf)(x) \stackrel{\Delta}{=} \int_a^b H(x - t) f(t)\, dt \qquad x \in [a, b] \tag{27}$$

[where H is the Heaviside function $H(x) = 1$ for $x \geq 0$, $H(x) = 0$ otherwise] the KDE $\tilde{f}_n$ can be shown to satisfy

$$\tilde{f}_n \stackrel{\Delta}{=} \arg \min_{f \in L_2([a,b])} (\|Lf - \tilde{F}_n\|_2^2 + \lambda\Omega(f)) \tag{28}$$

where $\tilde{F}_n(\bullet) \stackrel{\Delta}{=} \sum_{i=1}^n H(\bullet - x_i)$ is the *empirical distribution function* based on a random sample $t_n \stackrel{\Delta}{=} \{x_i\}_{i=1}^n$, $\|\bullet\|_2$ denotes the usual $L_2([a, b])$ norm and $\Omega: L_2([a, b]) \mapsto \mathbb{R}^+$ is a suitable regularizing functional. For example, $\Omega(f) = \|f\|_2^2$ yields a KDE with kernel function $K(x) \stackrel{\Delta}{=} \exp(-|x|/\sqrt{\lambda})/(2\sqrt{\lambda})$. While this theoretical link between the KDE and the regularization theory used in deriving the strict interpolation RBFN is interesting, it does not in itself suggest any significant new practical approaches to density estimation over the KDE. For this purpose, Vapnik (1982) develops an estimate of f based on the minimization of $\tilde{F}_n$ in place of F via the SRM method, as discussed previously.

The step from density estimation to regression estimation is a natural one. Suppose that the joint density of $(\boldsymbol{X}, Y)$ can be estimated with a KDE of the form

$$g_n(\boldsymbol{x}, y) \stackrel{\Delta}{=} \frac{1}{n} \sum_{i=1}^n \frac{1}{h_n^d} K\left(\frac{\boldsymbol{x} - \boldsymbol{X}_i}{h_n}\right) \cdot \frac{1}{h_n} K_1\left(\frac{y - Y_i}{h_n}\right) \tag{29}$$

where, for simplicity, we assume that $K(\boldsymbol{x}) \stackrel{\Delta}{=} \prod_{j=1}^d K_1(x_j)$, $\boldsymbol{x} = [x_j]_{j=1}^d$, is the product kernel formed from a valid one-dimensional kernel K_1. Then, using the approximate densities $\tilde{g}_n$ and $\tilde{f}_n$ in place of the true ones, we obtain the induced or *plug-in* regression estimate known as the *Nadaraya–Watson regression estimate (NWRE)* (Nadaraya, 1964, 1965; Watson, 1964)

$$\begin{aligned}
\mathbb{E}[Y|\boldsymbol{X} = \boldsymbol{x}] &\approx \int y \frac{\tilde{g}_n(\boldsymbol{x}, y)}{\tilde{f}_n(\boldsymbol{x})} dy \\
&= (\tilde{f}_n(\boldsymbol{x}))^{-1} \frac{1}{nh_n^d h_n} \sum_{i=1}^n \left\{ K\left(\frac{\boldsymbol{x} - \boldsymbol{X}_i}{h_n}\right) \int (h_n u + Y_i) K_1(u) h_n du \right\} \\
&= \frac{\sum_{i=1}^n Y_i K\left(\frac{\boldsymbol{x} - \boldsymbol{X}_i}{h_n}\right)}{\sum_{i=1}^n K\left(\frac{\boldsymbol{x} - \boldsymbol{X}_i}{h_n}\right)}
\end{aligned} \tag{30}$$

where we have used the facts

$$\int uK_1(u)du = 0 \qquad \int K_1(u)du = 1 \tag{31}$$

in the second line of the derivation and simplified to obtain the third line. It is not too hard to conjecture that the consistency of the density estimates $\tilde{g}_n$ and $\tilde{f}_n$ should carry over to the NWRE in Eq. (30). Indeed, Devroye (1981) showed that along with the basic KDE consistency conditions in Eq. (26), all that is needed for the $L_2(P_{T_n})$ consistency of $\tilde{f}_n$ with respect to f to hold a.e.–P_X is

$$\exists r, c_1, c_2 \in \mathbb{R}^+ : c_1 I_{\{\|x\| \leq r\}} \leq K(x) \leq c_2 I_{\{\|x\| \leq r\}} \tag{32}$$

that is, the kernel K is sufficiently smooth as to be bounded by two cylinders about the origin in $\mathbb{R}^d$.

For both the KDE and NWRE, so long as the kernel satisfies the necessary smoothness and boundedness conditions previously described, the exact shape of the kernel matters less to the resultant predictor performance than the correct choice of bandwidth h_n (Györfi et al., 1989; Bosq, 1996). Here, as with the ERM theory for ANNs, KRE theory offers several methods of selecting the bandwidth parameter according to data-based proxies for the desired performance or generalization measure which is typically quadratic. The difference, however, is that the procedures (as originally formulated) justify their choice of proxy by the criterion of *asymptotic optimality (a.o.)*. In what follows, we shall denote the performance measure by V and its proxy by $\tilde{V}$. Given T_n, let $\tilde{h}_n^*$ and $\tilde{f}_n^*$ (h_n^* and f_n^*) be the bandwidth and corresponding function estimate that minimize the proxy $\tilde{V}(\tilde{f}_n, T_n)$ [the actual generalization measure $V(f_n, f)$] over all possible estimates $\tilde{f}_n$ (f_n) derived from T_n. Then V a.o. means that

$$1 \leq \frac{V(\tilde{f}_n^*, f)}{V(f_n^*, f)} \xrightarrow{n\to\infty} 1 \tag{33}$$

where the convergence is at least i.p.–P_{T_n}, that is, we have weak V consistency. Although we shall have more to say regarding such asymptotically optimal data-based parameter selection procedures, it suffices here to note that they are typically computationally intensive. For example, for both the NWRE and KDE cases, one is often interested in the (squared-error) *loss* $\tilde{L}_2$ of an estimate $\tilde{f}_n$ with respect to the true function f given t_n, where

$$\tilde{L}_2(\tilde{f}_n, f, t_n) \triangleq \frac{1}{n}\sum_{i=1}^{n}(f(x_i) - \tilde{f}_n(x_i))^2 \tag{34}$$

and, in particular, the *expected loss* or *risk* $\tilde{R}_2$, that is,

$$\tilde{R}_2(\tilde{f}_n, f) \triangleq \mathbb{E}_{T_n}[\tilde{L}_2(\tilde{f}_n, f, T_n)] \tag{35}$$

To avoid confusion with the previously defined risk R_2 (which is independent of n), we shall also denote $\tilde{R}_2$ by $\tilde{R}_{2,n}$ when we wish to emphasize its dependence on n and call it the *MS fitting error (MSFE)*. One of the first proxies studied for the MSFE is the *ordinary* or *delete one cross-validation* (OCV or simply CV) function, computed in the case of a single bandwidth parameter h_n as

$$\mathrm{CV}(h_n, t_n) \triangleq \frac{1}{n}\sum_{i=1}^{n}(y_i - \tilde{f}_{-i,n}(x_i, h_n))^2 \tag{36}$$

where $\tilde{f}_{-i,n}(\bullet, h_n)$ is the function estimate specified by h_n and T_n with its ith training pair deleted. The CV proxy states that a reasonable estimate for h_n^*, the true MS fitting error-minimizing bandwidth parameter, should also yield a series of networks each of which is able to predict the one datum that was removed from T_n during its training. This proxy is known to have a variety of desirable theoretical properties, including asymptotic optimality in the case of noisy T_n with heteroskedastic errors (Andrews, 1991) and unbiased convergence to the global or unconditional MSE (Plutowski et al., 1994). It is also clear, however, that the minimization of $\mathrm{CV}(h_n, T_n)$ with respect to h_n generally involves the computation of n networks and their respective losses, which can be prohibitively complex. For linearly weighted networks such as the RBFN, however, the situation is somewhat better: it can be shown [for example, see Theorem 4.2.1 in (Wahba, 1990)] that the CV proxy can be equivalently stated as

$$\mathrm{CV}(h_n, t_n) = \frac{1}{n}\sum_{i=1}^{n}\left(\frac{y_i - \tilde{f}_n(x_i)}{1 - a_{ii}(h_n, t_n)}\right)^2 \tag{37}$$

where $\tilde{f}_n$ is the estimate corresponding to h_n and a_{ii} is the ith diagonal element of the *influence* or *hat matrix* $A(h_n, t_n)$, which by definition relates the estimated output vector $\tilde{y}$ to the training output vector y via $\tilde{y} \stackrel{\Delta}{=} A(h_n, t_n)y$. For example, in the case of the NWRE with bandwidth h_n, it is easily seen from Eq. (30) that $A(h_n, t_n) = \text{diag}^{-1}[k(x_i)^\top \mathbf{1}_n]_{i=1}^n K$, where $k(\bullet) \stackrel{\Delta}{=} [K((x - X_i)/h_n)]_{i=1}^n$, $\mathbf{1}_n$ is a vector of n-ones, and $K \stackrel{\Delta}{=} [k^\top(x_i)]_{i=1}^n$ is the $n \times n$ NWRE interpolation matrix.

It is apparent in Eq. (37) that the CV criterion is not invariant to rotations of the co-ordinate axes for noisy T_n, that is, applying an orthogonal transformation to the problem Eq. (17). The desire for such rotational invariance motivated the introduction of the so-called *generalized cross-validation (GCV)* proxy (Craven and Wahbe, 1979; Wahba, 1990). In the following, let h_n be a generic parameter to be estimated from a realized example set t_n. Then the GCV proxy is defined as

$$\begin{aligned} \text{GCV}(h_n, t_n) &= \tilde{J}_2(\tilde{f}_n, t_n)/(1 - n^{-1}\,\text{tr}\,A(h_n, t_n))^2 \\ &= \frac{(1/n)\|[I - A(h_n, t_n)]y\|^2}{\{(1/n)\,\text{tr}[I - A(h_n, t_n)]\}^2} \end{aligned} \tag{38}$$

For the strict interpolation RBFN with regularization parameter λ_n, $\tilde{y} = Gw$ so $A(\lambda_n, t_n) = G(G + \lambda_n I)^{-1}$, hence GCV can be seen as an "averaged" version of CV in which the a_{ii} have been replaced by their average value $\sum_{i=1}^n a_{ii}/n = \text{tr}\,A(h_n, t_n)/n$. Like the CV proxy, the GCV proxy is known to have several favorable theoretical properties such as asymptotic optimality, which we shall discuss at greater length in Section 1.4.1.

As mentioned earlier, for stochastic T_n and risk measured by the usual (global) MSE, the underlying function f is precisely the conditional mean of the output RV Y given the input RV X. Comparing the NWRE with a corresponding strict interpolation RBFN under these conditions, we find that the NWRE appears to be a strict interpolation RBFN with norm-weighting matrix I/h_n for all basis functions, as in the KDE case, but with the weights w_i now dependent upon both T_n (through the output training values Y_i) and the evaluation point x, viz.

$$w_i = w_i(x; T_n) \stackrel{\Delta}{=} \frac{Y_i}{\sum_{i=1}^n K\left(\dfrac{x - X_i}{h_n}\right)} \qquad \tilde{f}_n(x) = \sum_{i=1}^n w_i(x; T_n) K\left(\frac{x - X_i}{h_n}\right) \tag{39}$$

This form for the weights suggests that other such general weighting schemes may also be appropriate. By expressing the NWRE in a slightly different form as a *weighted output* mean

$$\tilde{f}_n(\bullet) \stackrel{\Delta}{=} \sum_{i=1}^{n} a_{n,i}(\bullet; T_n)Y_i \tag{40}$$

Stone (1977) gave a set of sufficient conditions on the output weight vector function $\boldsymbol{a}_n \stackrel{\Delta}{=} [a_{n,i}]_{i=1}^n$ for the $L_2(P_{XY} \times P_{T_n})$ consistency of $\tilde{f}_n$ with respect f, of which the following two are necessary:

$$\sum_{i=1}^{n} a_{n,i}(\boldsymbol{X}) \overset{n\to\infty}{\to} 1 \text{ i.p.} \qquad \max_{i=1,2,\ldots,n} |a_{n,i}(\boldsymbol{X})| \overset{n\to\infty}{\to} 0 \text{ i.p.} \tag{41}$$

These conditions become both necessary and sufficient when $\boldsymbol{a}_n$ is a sequence of *probability weights*, that is, $\sum_{i=1}^n a_{n,i}(\boldsymbol{x}) = 1$ and $a_{n,i}(\boldsymbol{x}) \geq 0$ for all $\boldsymbol{x} \in \mathbb{R}^d$. Unfortunately, the strict interpolation RBFN *effective* output weights

$$\boldsymbol{a}_n(\bullet) = (\boldsymbol{G} + \lambda \boldsymbol{I})^{-1}\boldsymbol{g}(\bullet) \qquad \boldsymbol{g}(\bullet) \stackrel{\Delta}{=} [G_i(\bullet)]_{i=1}^n \tag{42}$$

do not satisfy all the sufficient conditions. At the same time, verifying the two necessary conditions is difficult, since the output weight sequence depends on the exact choices of the corresponding sequences for λ_n, the regularization parameter, and $\boldsymbol{U}_n$, the norm weighting matrix. Another route to consistency for the strict interpolation RBFN that indicates how these sequences should be chosen is highly desirable.

Thus far, the results for KRE have been stated for independent samples $(\boldsymbol{X}_i, Y_i) \in T_n$ and evaluation point $(\boldsymbol{X}, Y)$. For the NWRE, however, many of the same consistency results have been shown to hold with minor changes when the samples in T_N are drawn from a process whose dependence structure is described by *mixing conditions* (Douhkan, 1994; Bradley, 1986). In this respect, the application of KRE to correlated data as would occur in time-series analysis rests upon a sounder foundation than the current ERM theory for ANN. We shall see shortly how this body of theory can be exploited in the design of RBFN for similar applications.

Limitations of the Kernel Regression Approach

Appealing as the solid theoretical basis for the KRE and (in particular) the NWRE design procedures may be, there are some limitations to this approach. As a practical example, the KRE/NWRE framework does not account for the effects of selecting $n < N$ basis function centers from the N available data in T_N, a situation handled by the *penalized least-squares (PLS)* theory (Golitschek and Schumaker, 1990) supporting regularized RBFNs.[16] More significant, however, is the fact that the strict interpolation RBFN design procedure does not yield a set of output weights that fits clearly into the KRE/NWRE framework developed by Stone (1977), as previously explained. One obvious remedy to this difficulty is to use *normalized RBFN* (Xu et al., 1994), that is,

$$\tilde{f}_n(\bullet) = \frac{\sum_{i=1}^{n} w_i G_i(\bullet)}{\sum_{i=1}^{n} G_i(\bullet)} \tag{43}$$

where the norm-weighting matrix $\boldsymbol{U}_n = \boldsymbol{I}/h_n$, with $\{h_n\}$ satisfying Eq. (26) as for the KDE and NWRE. In the analysis of Xu et al. (1994), the centers are chosen as per the *random centers* method of Broomhead and Lowe (1988) previously described and, correspondingly, the weights $[w_i]_{i=1}^n$ solve the least-squares pseudo-inverse equation given in Eq. (10) with the modified interpolation matrix $\boldsymbol{G}'$ in place of $\boldsymbol{G}$, where

$$[\boldsymbol{G}']_{ij} = \frac{[\boldsymbol{G}]_{ij}}{\sum_{k=1}^{n} [\boldsymbol{G}]_{kj}} \tag{44}$$

From the clear similarity to the NWRE, it is expected that such normalized RBFNs should be consistent in the same modes that the NWRE is. While the rigorous proofs of Xu et al. (1994) supporting this intuitive notion are useful, this approach fails to include the important case of regularized strict interpolation RBFNs with which we are primarily concerned. One advantage of explicit regularization for strict interpolation RBFNs is that computationally efficient selection procedures for λ exist that are asymptotically optimal in the *actual* loss for a given training input sequence rather than just the *average* loss or risk (see the references for the scenarios listed in Section 1.4.1); in this sense, the *pointwise* behavior

[16]See note 14 in Appendix A.1.

of the regularized strict interpolation RBFN estimate can be proven correct.

GUIDED TOUR OF BOOK

Given that the current ANN and KRE/NWRE approaches to RBFN design leave something to be desired, the object of this book is to demonstrate that a rigorous basis for regularized strict interpolation RBFN can be constructed from the related areas of *regularized* or *penalized least-squares* (PLS) fitting and *spline smoothing*. Building upon this basis, it is possible to:

A1. Show that for stochastic T_n and performance measured by the MSFE $\tilde{R}_2$, the ANN ERM method is nonoptimal because it corresponds to the regularization parameter sequence with $\lambda_n = 0$ for all n, whereas the optimal regularization parameter λ_n^* for each n is nonzero.

A2. Give an explicit method for the computation of asymptotically optimal estimates $\tilde{\lambda}_n$ of the MSFE minimizing regularization parameter $\tilde{\lambda}_n^*$.

A3. Show that the MSFE $\tilde{R}_2$ converges a.s. to the (global) MSE or risk R_2.

A4. Prove the mean square consistency of the regularized strict interpolation RBFN with asymptotically optimal regularization parameter sequence under the same conditions as is known for the NWRE.

A5. Show that a strict interpolation RBFN with positive-definite kernel and designed by the ERM method without regularization, that is with $\lambda_n = 0$ for all n, cannot be mean square consistent.

Items A1 and A2 are discussed in the PLS literature, for example, see Golitschek and Schumaker (1990) and Wahba (1990), while item A5 is a direct consequence of the PLS theory as shown later in Section 1.5. Result A4, however, requires more work. First, we present a *constructive* proof of the a.s. uniform and mean-square approximation of the NWRE over compact sets with arbitrary rates of convergence by a class of suitably regularized strict interpolation RBFN for the case of stochastic T_n, where the (X_i, Y_i) are drawn from either an i.i.d. or mixing process with a stationary marginal density for $\{X_i\}$. This *asymptotic equivalence* between the NWRE and the strict interpolation RBFN justifies the application of the large body of theory surrounding NWRE design and motivates the derivation of result A4 by way of comparison. This key result can be established after result A3 is established to link asymptotically the global risk R_2 to its proxy, the MSFE $\tilde{R}_2$.

Once these theoretical tools are in place, they find natural applications in the areas of probability estimation, classification, and time-series prediction. For the first two related areas, we can:

B1. Prove the mean-square consistency of regularized strict interpolation RBFN for posterior probability estimation when using asymptotically optimal regularization parameter sequence.

B2. Show the Bayes risk consistency of the approximate Bayes decision rules based on such regularized strict interpolation RBFN posterior probability estimates.

B3. Prove that mean-square consistency of posterior probability estimates implies weak convergence of the corresponding approximate Bayes decision rules and their respective classifier error rates (consistent or not). Thus positivity and normalization of mean-square consistent posterior probability estimates are not required for Bayes risk consistency.

Not surprisingly, the flexibility of RBFNs can be useful for modeling nonlinear time series. As a first step in this direction, we consider the generalization from the usual linear autoregressive (AR) time series to the class of Markovian *nonlinear AR (NLAR)* time series generated by an i.i.d. noise process (see Section 3.5 for details). For the prediction of this class, we can:

C1. Show that, with suitable modifications, result A3 carries over to the NLAR case. This result justifies the use of data-based parameter estimation procedures such as GCV for the asymptotically optimal selection of λ in such time series.

C2. Derive recursive updating algorithms for regularized strict interpolation RBFN similar to the linear RLS updates for *one-step-ahead (OSA)* prediction when new training data are continually available. The following two basic types of network updating are considered:

   (a) A new weight/basis function associated with the most recently available training data is added per update (*augmented* or *infinite memory* case).

(b) The weight/basis function associated with the oldest available training data is discarded for a new weight/basis function associated with the most recently available training data (*fixed-size* or *finite memory* case).

C3. Characterize the performance gain in using nonlinear autoregressive modeling over the usual linear autoregressive modeling via experiments on the OSA prediction of speech signals.

C4. Show experimentally the improvement possible by linearly combining several predictor outputs using the exponentially weighted RLS algorithm.

We close the book with a few concluding remarks which offer some perspective on the contributions and limitations of the work, the latter naturally leading to some problems open for investigation.

1 BASIC TOOLS

1.1 INTRODUCTION

To achieve a deeper understanding of strict interpolation radial basis functions (RBFNs) for our intended applications, we need to develop a body of theory describing the basic properties and behavior of strict interpolation RBFNs. Many approaches are, of course, possible but our chosen approach of relating the strict interpolation RBFN to the classical Nadaraya–Watson regression estimate (NWRE) is perhaps among the simplest and most accessible. We do so in several stages: First we bound *deterministically* the pointwise difference between the two estimators for any given training set, showing that this difference decreases polynomially as a function of n, the size of the training set. The proof of the bound specifies how the strict interpolation RBFN is to be designed to realize the bound, that is, the proof is *constructive*. While this result is interesting, most practical applications involve stochastic or random data in some way, hence we show how this result can be extended to hold uniformly in the (a) pointwise almost sure and (b) mean-square senses. These *probabilistic* approximation results relating strict interpolation RBFNs to the NWRE hold only for a special class of RBFNs characterized by an input-dependent regularization parameter sequence growing to infinity at some minimum rate. Despite this limitation, we then show how the trivial consistency of this special class (deduced by comparison with the NWRE) and the notion of asymptotically optimal regularization parameter sequence can be exploited to prove the mean-square (MS) consistency of the more usual regularized strict interpolation RBFNs when designed with such asymptotically optimal regularization parameter sequences. The main tool we use to link these two ideas is a proof of the convergence of the MS fitting error (MSFE) $\tilde{R}_2$ (in which the regularization parameter sequence is chosen to be optimal) to the true global risk R_2, that is, the global mean-square error (MSE). Then by a comparison argument, we conclude that a regularized strict interpolation RBFN with asymptotically optimal regularization parameter sequence cannot have global MSE greater than that of any corresponding strict interpolation RBFN with the special choice of input-dependent regularization parameter sequence. Since the latter is known to be mean-square consistent (whenever the corresponding NWRE is), it follows that regularized strict interpolation RBFNs with asymptotically optimal regularization parameter sequence must also be mean-square consistent under the same conditions.

1.2 ASYMPTOTIC EQUIVALENCE OF NWRE TO REGULARIZED STRICT INTERPOLATION RBFN VIA CONSTRUCTIVE APPROXIMATION

At this point it would be useful to discuss the basic results upon which the forthcoming applications are based. We begin by showing that any NWRE $\tilde{f}_n'$ with radial kernel K' designed from a given training set t_n can be approximated with vanishing error at a given point $z \in \mathbb{R}^d$ by a suitably designed regularized strict interpolation RBFN $\tilde{f}_n$. In the following, the primed quantities refer to constructions involving the NWRE with the non-normalized kernel K' as defined in the preamble to the lemma.

Lemma 1. *Let $\tilde{f}_n'$ be a NWRE with radial kernel $K' = C \cdot K$, where $C \stackrel{\Delta}{=} \sup_{z \in \mathbb{R}^d} K'(z)$, and bandwidth parameter h_n designed from a given training set t_n. Then for any $n > 1$, $\alpha > \log(C/(h_n^d \tilde{p}_n(z)))/\log n$, and $z \in \mathbb{R}^d$ such that the denominator $nh_n^d \tilde{p}_n(z) \stackrel{\Delta}{=} \mathbf{1}^\top \mathbf{g}_n'(z)$ of $\tilde{f}_n'$ is not zero, a regularized strict interpolation RBFN $\tilde{f}_n$ with kernel K may be constructed such that*

$$|\tilde{f}_n(z) - \tilde{f}_n'(z)| \leq \frac{C^2 M}{n^{\alpha/2} h_n^d \tilde{p}_n(z)[n^{\alpha/2} h_n^d \tilde{p}_n(z) - Cn^{-\alpha/2}]} \tag{1.1}$$

where $\mathbf{y} \leq M$ and where $\mathbf{1}_n$ is a constant vector of n ones.

As an aside, the identification of the NWRE denominator $\mathbf{1}^\top \boldsymbol{g}_n'(z)$ with the *Parzen window (density) estimate (PWE)* (Parzen, 1962) $\tilde{p}_n$ of the marginal input density p is merely suggestive of a decomposition that is useful in subsequent proofs.

Proof. For the regularized strict interpolation RBFN, set $\boldsymbol{U}_n \stackrel{\Delta}{=} \boldsymbol{I}/h_n$ and $\lambda_n = \lambda_n(z) = n^\alpha \boldsymbol{g}_n^\top(z)\mathbf{1}_n = n^{\alpha+1}h_n^d\tilde{p}_n(z)/C$, where $\alpha \geq 0$ is an exponent to be determined later. Let $\boldsymbol{y}_n' \stackrel{\Delta}{=} n^\alpha \boldsymbol{y}_n$ so that

$$\tilde{f}_{n,\infty}'(z) = \boldsymbol{g}_n^\top(z)(\lambda_n(z)\boldsymbol{I})^{-1}\boldsymbol{y}_n' \tag{1.2}$$

Comparing the NWRE output to that of the regularized strict interpolation RBFN designed from t_n, we find that the difference can be bounded (by the Cauchy–Schwarz inequality) as

$$\begin{aligned}
|\tilde{f}_n(z) - \tilde{f}_n'(z)| &= |\langle \boldsymbol{g}_n(z), (\boldsymbol{G}_n + \lambda_n(z)\boldsymbol{I})^{-1}\boldsymbol{y}_n' - (\lambda_n(z)\boldsymbol{I})^{-1}\boldsymbol{y}_n'\rangle| \\
&\leq \|\boldsymbol{g}_n(z)\| \|(\boldsymbol{G}_n + \lambda_n(z)\boldsymbol{I})^{-1} - (\lambda_n(z)\boldsymbol{I})^{-1}\| \|\boldsymbol{y}_n'\| \\
&\leq \|\boldsymbol{g}_n(z)\| \frac{\|\boldsymbol{G}_n\|\|\boldsymbol{I}\|}{\lambda_n(z)(\lambda_n(z) - \|\boldsymbol{G}_n\|)} \|\boldsymbol{y}_n'\| \quad \text{for} \quad \lambda_n(z) > \|\boldsymbol{G}_n\|
\end{aligned} \tag{1.3}$$

Using the Euclidean norm as an upper bound for all quantities except for $\boldsymbol{G}_n$, which we bound in Fröbenius norm[1] as $\|\boldsymbol{G}_n\| \leq n$, we obtain

$$\begin{aligned}
|\tilde{f}_n(z) - \tilde{f}_n'(z)| &\leq \sqrt{n}\frac{n}{\lambda_n(z)(\lambda_n(z) - n)} n^\alpha \sqrt{n}M \\
&\leq \frac{n^2 n^\alpha M}{\lambda_n(z)(\lambda_n(z) - n)}
\end{aligned}$$

For our choice of $\lambda_n(z)$, this latter result can be rewritten as

$$\begin{aligned}
|\tilde{f}_n(z) - \tilde{f}_n'(z)| &\leq \frac{n^{\alpha+2}C^2 M}{n^{\alpha+1}h_n^d\tilde{p}_n(z)(n^{\alpha+1}h_n^d\tilde{p}_n(z) - nC)} \\
&\leq \frac{C^2 M}{n^{\alpha/2}h_n^d\tilde{p}_n(z)[n^{\alpha/2}h_n^d\tilde{p}_n(z) - Cn^{-\alpha/2}]}
\end{aligned} \tag{1.4}$$

[1]See note 1 in Appendix A.2.

The condition on $\lambda_n(z)$ in Eq. (1.3) can be satisfied by choosing

$$\lambda_n(z) > n \Rightarrow n^{\alpha+1} h_n^d \tilde{p}_n(z)/C > n \Rightarrow \alpha > \log(C/(h_n^d \tilde{p}_n(z)))/\log n \quad (1.5)$$

□

The proof of the lemma shows that by selecting the strict interpolation RBFN regularization parameter λ_n to be the denominator of the NWRE and scaling it along with training outputs in t_n at rate n^α, an arbitrarily fast rate of approximation at a *fixed point* $z \in \mathbb{R}^d$ is possible. We would, of course, like to extend this pointwise approximation result to corresponding a.s. uniform and MS approximation results in the case of stochastic T_n. As a direct route to such results, we shall consider a special class F_z of regularized strict interpolation RBFNs in which λ_n (and hence $\mathbf{w}_n$) is permitted to vary with its input $z \in \mathbb{R}^d$ as in the proof of the lemma. This class is therefore a slight generalization of the usual class of regularized strict interpolation RBFNs in which λ_n (and hence $\tilde{\mathbf{w}}_n$) is set once via t_n for all inputs z. As will be explained further on, the generalization does not affect the overall tenor of the results on the relative suboptimality of NWREs with respect to the risk or MSFE over T_n.

In view of applications to time-series analysis considered in later chapters, we shall indicate (discrete) time dependence by parenthesized indices instead of subscripts and denote the input–output processes by $\{\mathbf{Z}(i)\}$ and $\{Y(i)\}$, respectively. The upper bound of Lemma 1 can be generalized to hold a.s. uniformly and in MS over a compact set $D \subset \mathbb{R}^d$ instead of a single point by assuming conditions that ensure that $\tilde{p}_n$ is lower bounded in the respective modes. For convenience, we assume the sufficient condition that the input processes have a common marginal density[2] p to which the Parzen window density estimate $\tilde{p}_n$ defined in the preamble of Lemma 1 converges (in the same mode). When this situation is not applicable, for example, when the input process is nonstationary, a more general condition that allows the required lower bounding is condition (A.1) from Györfi et al. (1989)

$$\begin{array}{ll} \exists \Gamma < \infty \text{ such that } \forall i \in \mathbb{N} \text{ and } \forall B \in \mathcal{B}(\mathbb{R}^d) & \Pr(\mathbf{Z}(i) \in B) \leq \Gamma \mu(B) \\ \exists \gamma, \epsilon > 0 \text{ such that } \forall i \in \mathbb{N} \text{ and } \forall B \in \mathcal{B}(D_\epsilon) & \Pr(\mathbf{Z}(i) \in B) \geq \gamma \mu(B) \end{array} \quad (A.1)$$

[2]See note 2 in Appendix A.2.

where $\mathcal{B}(\mathbb{R}^d)$ [resp. $\mathcal{B}(D_\epsilon)$] is the σ-algebra of the Borel sets on $\mathbb{R}^d$ (resp. on D_ϵ), μ is the Lebesgue measure on $\mathbb{R}^d$, and $D_\epsilon \ni D$ is the set of all ϵ-neighborhood compacts[3] of D. Roughly speaking, the first condition of (A.1) states that all the marginal densities of the process $\{\boldsymbol{Z}(i)\}$ are (at least) piecewise continuous and bounded over the open sets of $\mathbb{R}^d$. Similarly, the second condition of (A.1) states that all the marginal densities of $\{\boldsymbol{Z}(i)\}$ are lower bounded away from zero on any compact set which is ϵ close to the region of interest D. As mentioned in Remark 3.3.1 of Györfi et al. (1989), these conditions are sufficient to ensure that the Parzen window density estimate $\tilde{p}_n$ does not vanish on D for each n. As an aside, since $\{\tilde{p}_n\}$ is a sequence of positive, continuous functions converging (in one of the aforementioned modes) to p over a compact set D, this result also implies that p must be lower bounded away from zero over D.

Theorem 1. *Assume that $\{\boldsymbol{Z}(i)\}$ has a stationary marginal measure P and density p with respect to Lebesgue measure. Let D be a compact subset of $\mathbb{R}^d$ with $p(z) > 0$ for all $z \in D$. Then*

1. *If $|Y(i)| < M$ almost surely (a.s.) for all i and if K', $\{h_n\}$, and p are such that*

$$\sup_{z \in D} |\tilde{p}_n(z) - p(z)| \text{ a.s.}-P_{T_n} \overset{n\to\infty}{\to} 0 \tag{1.6}$$

then $\exists N = N(p, D, K', \{h_n\})$ such that for any $n > N$ and $\alpha > \max[2, \log(2C/(h_n^d m))/\log n]$, a regularized strict interpolation RBFN $\tilde{f}_{n,\infty} \in F_z$ may be constructed such that

$$\sup_{z \in D} |\tilde{f}_{n,\infty}(z) - \tilde{f}'_n(z)| = \mathcal{O}(C^2 M n^{-\alpha} h_n^{-2d} m^{-2}) \text{ a.s.}-P_{T_n} \tag{1.7}$$

where $m = m(D) \overset{\Delta}{=} \inf_{z \in D} p(z)$.

2. *If $\mathbb{E}[Y^2(i)] < M^2$ for all i and if K', $\{h_n\}$, and p are such that*

$$\sup_{z \in D} \mathbb{E}_{T_n}[|\tilde{p}_n(z) - p(z)|^2] \overset{n\to\infty}{\to} 0 \tag{1.8}$$

[3]See note 3 in Appendix A.2.

and there exist positive constants R_1, R_2, R_3 and ν such that

$$\lim_{n\to\infty} \sup_{z\in D} (|\tilde{f}_{n,\infty}(z)| + |\tilde{f}_n'(z)|) < R_1 \ a.s.\text{–}P_{T_n} \tag{1.9}$$

$$\lim_{n\to\infty} \sup_{\substack{i,j=1,\dots,n \\ (x,y)\in\mathbb{R}^d\times\mathbb{R}^d}} \frac{p_{ij}(x,y)}{p(x)p(y)} < R_2 \tag{1.10}$$

$$\lim_{n\to\infty} \inf_{z\in D} n^{\nu} n h_n^d \tilde{p}(z) > R_3 > 0 \ a.s.\text{–}P_{T_n} \tag{1.11}$$

where $p_{ij}(\bullet,\bullet)\colon \mathbb{R}^d \times \mathbb{R}^d \to \mathbb{R}^+$ is the joint density for $\boldsymbol{Z}(i)$ and $\boldsymbol{Z}(j)$, then $\exists N = N(p, D, K', \{h_n\})$ such that for $n > N$ and $\alpha > \max(1, \nu, \log(2C/(h_n^d m))/\log n)$, a regularized strict interpolation RBFN $\tilde{f}_{n,\infty} \in F_z$ may be constructed such that

$$\sup_{z\in D} \mathbb{E}_{T_n}[|\tilde{f}_{n,\infty}(z) - \tilde{f}_n'(z)|^2] = \mathcal{O}(C^2 M \sqrt{L} \|p\|_2 \|K\|_2^2 n^{-\alpha} h_n^{-d} m^{-2}) \tag{1.12}$$

where m is as before and $L = L(D) \triangleq \sup_{z\in D} p(z)$.

Proof. We treat each case separately:

1. It is easy to show that Eq. (1.6) implies that by choosing N to satisfy

$$n > N \Rightarrow \sup_{z\in D} |\tilde{p}_n(z) - p(z)| < \frac{m}{2} \tag{1.13}$$

we have $n > N \Rightarrow \tilde{p}_n(z) \geq m/2$ for all $z \in D$. Hence for $n > N$, we may replace $\tilde{p}_n(z)$ with $m/2$ in the denominator of the upper bound, and the term $Cn^{-\alpha/2}$ can be dominated by selecting a sufficiently large constant to multiply the numerator of the order bound, that is, $\exists L > 0$ such that for $n > N$,

$$\frac{L \cdot C^2 M}{n^{\alpha} h_n^{2d} m^2} > \frac{C^2 M}{n^{\alpha/2} h_n^d (m/2)[n^{\alpha/2} h_n^d m/2 - Cn^{-\alpha/2}]} \tag{1.14}$$

From the basic kernel density estimate (KDE) consistency condition $nh_n^d \xrightarrow{n\to\infty} \infty$, requiring $\alpha > \max(2, \log(2C/(h_n^d m))/\log n)$ ensures that the approximation error vanishes with increasing n.

2. While the convergence rates for this case must be at least as rapid as for the a.s. uniform case [by squaring and taking expectations on

both sides of Eq. (1.3) before computing the supremum on the left-hand side], we can obtain slightly better convergence rates with tighter MS estimates of the terms in Eq. (1.3). We begin by noting that it is sufficient to demonstrate the corresponding result in absolute value, since

$$\begin{aligned}\mathbb{E}_{T_n}[|\tilde{f}_{n,\infty}(z) - \tilde{f}_n'(z)|^2] &= \mathbb{E}_{T_n}[(\tilde{f}_{n,\infty}(z) - \tilde{f}_n'(z))(\tilde{f}_{n,\infty}(z) + \tilde{f}_n'(z)) \\ &\quad + 2\tilde{f}_n'(z)(\tilde{f}_n'(z) - \tilde{f}_{n,\infty}(z))] \\ &\leq \sup_{z \in D}(|\tilde{f}_{n,\infty}(z) + \tilde{f}_n'(z)| + 2|\tilde{f}_n'(z)|) \\ &\quad \cdot \mathbb{E}_{T_n}[|\tilde{f}_{n,\infty}(z) - \tilde{f}_n'(z)|] \end{aligned} \tag{1.15}$$

where the supremum is $\mathcal{O}(R_1)$ for n sufficiently large by assumption Eq. (1.9). Returning to the expectation term, taking expectations with respect to P_{T_n} on both sides of Eq. (1.3) and applying Cauchy–Schwarz gives

$$\begin{aligned}\mathbb{E}_{T_n}[|\tilde{f}_{n,\infty}(z) - \tilde{f}_n'(z)|] \leq &\mathbb{E}_{T_n}^{1/2}[\|\boldsymbol{g}_n(z)\|^2]\mathbb{E}_{T_n}^{1/2}[\|\boldsymbol{G}_n\|^2] \\ &\cdot \mathbb{E}_{T_n}^{1/2}[\lambda_n^{-2}(z)(\lambda_n(z) - \|\boldsymbol{G}_n\|)^{-2}]\mathbb{E}_{T_n}^{1/2}[\|\boldsymbol{y}_n'\|^2]\end{aligned} \tag{1.16}$$

The first term squared $\mathbb{E}_{T_n}[\|\boldsymbol{g}_n(z)\|^2]$ can be asymptotically bounded in Euclidean norm as

$$\begin{aligned}\mathbb{E}_{T_n}[\|\boldsymbol{g}_n(z)\|^2] &= \sum_{i=1}^{n} \mathbb{E}_{T_n}\left[K^2\left(\frac{z - \boldsymbol{Z}(i)}{h_n}\right)\right] \\ &= \mathcal{O}(nh_n^d p(z)\|K\|_2^2) \text{ a.e.}\end{aligned} \tag{1.17}$$

where $\|\bullet\|_2$ is the standard L_2 norm with respect to Lebesgue measure, and we have used the fact that [see Eq. (2.10) in Bosq (1996)]

$$\left|\int_{\mathbb{R}^d} K^2\left(\frac{\boldsymbol{x} - \boldsymbol{u}}{h_n}\right)p(\boldsymbol{u})\, d\boldsymbol{u} - h_n^d p(\boldsymbol{x})\|K\|_2^2\right| \overset{n\to\infty}{\rightarrow} 0 \text{ a.e.} \tag{1.18}$$

where $\|\bullet\|_2$ is the usual $L_2(\mathbb{R}^d)$ norm with respect to Lebesgue measure. Similarly, we bound the second term squared, $\mathbb{E}_{T_n}[\|\boldsymbol{G}_n\|^2]$ in Fröbenius norm and apply Eq. (1.18) with Lebesgue-dominated convergence[4] to obtain

$$\mathbb{E}_{T_n}[\|\boldsymbol{G}_n\|^2] \leq \sum_{i,j=1}^{n} \mathbb{E}_{\boldsymbol{Z}(i),\boldsymbol{Z}(j)}\left[K^2\left(\frac{\boldsymbol{Z}(i)-\boldsymbol{Z}(j)}{h_n}\right)\right]$$
$$= \mathcal{O}\left(n^2 \int_{\mathbb{R}^d}\int_{\mathbb{R}^d} K^2\left(\frac{z(i)-z(j)}{h_n}\right) p(z(i))p(z(j))dz(i)dz(j)\right)$$
$$= \mathcal{O}(n^2 h_n^d \|p\|_2^2 \|K\|_2^2) \tag{1.19}$$

where we have applied Eq. (1.10). For the square of the middle term, we may again apply the majorization $\|\boldsymbol{G}_n\| < n$ and use the same argument as for Eq. (1.14) to obtain the estimate

$$\mathbb{E}_{T_n}[\lambda_n^{-2}(z)(\lambda_n(z) - \|\boldsymbol{G}_n\|)^{-2}] \leq L \cdot \mathbb{E}_{T_n}[\lambda_n^{-4}(z)]$$
$$\forall z \in D \quad \text{when} \quad n > N_1 \tag{1.20}$$

for some $L > 0$ and $N_1 \in \mathbb{N}$. Next, we may substitute p for $\tilde{p}_n$ in the expectation with error bounded by

$$|\mathbb{E}_{T_n}[\lambda_n^{-4}(z)] - \lambda^{-4}(z)|$$
$$\leq \mathbb{E}_{T_n}[|\lambda_n^{-4}(z) - \lambda^{-4}(z)|]$$
$$\leq \mathbb{E}_{T_n}\left[\left|\frac{\lambda_n^4(z) - \lambda^4(z)}{\lambda_n^4(z)\cdot\lambda^4(z)}\right|\right]$$
$$\leq \mathbb{E}_{T_n}\left[\left|\frac{(\lambda_n^2(z) + \lambda^2(z))(\lambda_n(z) + \lambda z))(\lambda_n(z) - \lambda z))}{\lambda_n^4(z)\cdot\lambda^4(z)}\right|\right] \tag{1.21}$$

where $\lambda(z) \triangleq n^{\alpha+1} h_n^d p(z)/C$, whence, by the Cauchy–Schwarz inequality,

$$\sup_{z\in D} |\mathbb{E}_{T_n}[\lambda_n^{-4}(z)] - \lambda^{-4}(z)|$$
$$\leq \sup_{z\in D} \mathbb{E}_{T_n}^{1/2}\left[\left|\frac{(\tilde{p}_n^2(z) + p^2(z))(\tilde{p}_n(z) + p(z))}{(n^{\alpha+1}h_n^d\tilde{p}_n(z)/C)^4 \cdot p^4(z)}\right|^2\right] \tag{1.22}$$
$$\cdot \sqrt{\sup_{z\in D} \mathbb{E}_{T_n}[|\tilde{p}_n(z) - p(z)|^2]} \tag{1.23}$$

[4]See note 4 in Appendix A.2.

By conditions Eq. (1.9) and Eq. (1.11), the first sup term is (at least) bounded for α sufficiently large, while the second term vanishes by Eq. (1.8). Therefore we have that, for n sufficiently large,

$$\sup_{z\in D} \mathbb{E}_{T_n}[\lambda_n^{-2}(z)(\lambda_n(z) - \|G_n\|)^{-2}]$$
$$= \mathcal{O}\left(\sup_{z\in D} p^{-4}(z)\right) = \mathcal{O}((n^{\alpha+1}h_n^d m/C)^{-4}) \qquad (1.24)$$

The square of the last term $\mathbb{E}_{T_n}[\|y_n\|^2]$ is bounded (by assumption) by $nn^{2\alpha}M^2$, so that combining the square roots of the previous terms leaves us with the conclusion that for sufficiently large n,

$$\sup_{z\in D} \mathbb{E}_{T_n}[|\tilde{f}_{n,\infty}(z) - \tilde{f}_n'(z)|^2]$$
$$= \mathcal{O}\left(\frac{(nh_n^d L\|K\|_2^2)^{1/2} \cdot (n^2 h_n^d \|p\|_2^2 \|K\|_2^2)^{1/2} \cdot \sqrt{n}n^\alpha M}{(n^{\alpha+1}h_n^d m/C)^2}\right)$$
$$= \mathcal{O}(C^2 M\sqrt{L}\|p\|_2\|K\|_2^2 n^{-\alpha} h_n^{-d} m^{-2}) \qquad (1.25)$$

In order for Eq. (1.16) to be valid, we require that Eq. (1.3) hold uniformly over D, leading to the previous condition of $\alpha > \log(2C/(h_n^d m))/\log n$, where m is as defined in Eq. (1.13). As an aside, if Eq. (1.8) is weakened to the corresponding mean-input case, then Eq. (1.3) need only hold a.s.–P. Finally, the lower limits of 1 and ν imposed on α by the maximum function ensures that the upper bound on the MS approximation error in Eq. (1.12) decreases as $n \to \infty$ by the consistency condition $nh_n^d \overset{n\to\infty}{\longrightarrow} \infty$. □

One may find in the literature numerous sets of conditions under which Eq. (1.6) and Eq. (1.8) hold. In particular, we refer the reader to Lemma 2.1, Theorem 2.2, and Corollary 2.2 in the case of Eq. (1.6), and Theorem 2.1 and Corollary 2.1 in the case of Eq. (1.8), all from Bosq (1996). Rather than repeat the theorems and proofs verbatim, we merely make a few pertinent observations concerning their application and consequences:

1. The quoted results from Bosq (1996) hold for any kernel satisfying Eqs. (24) and (25) in the Introduction, including:
   (a) The *histogram* or *rectangular* kernel $K(\bullet) \overset{\Delta}{=} I_{[-1/2,+1/2]^d}(\bullet)$.
   (b) The *Gaussian* kernel $K(\bullet) \overset{\Delta}{=} (2\pi)^{-d/2}\exp(-\|\bullet\|^2/2)$.

(c) The *Epanechnikov* kernel $K(\boldsymbol{x}) \triangleq (\frac{3}{4}\sqrt{5})^d \prod_{i=1}^d (1 - x_i^2/5) I_{[-\sqrt{5},+\sqrt{5}]}(x_i)$, $\boldsymbol{x} \in \mathbb{R}^d$. This multidimensional quadratic product kernel can be shown to be optimal in the sense of minimizing the asymptotic MS density estimation error while having compact support (Epanechnikov, 1969).

Although conditions (24) and (25) in the Introduction are commonly assumed in both kernel density and regression estimation to ensure consistency, one should be aware that kernels that do not satisfy these conditions can have other desirable properties. Specifically, in the one-dimensional i.i.d. case, studies exist which show that KDEs based on strictly positive kernels cannot be globally unbiased for any finite training set and have limited asymptotic rate of bias convergence (Rosenblatt, 1956; Hand, 1982). From these results, Lowe (1995) argues for the use of *nonpositive* (definite) kernels for general RBFNs, that is, those not strictly of the regularization type. The same work also indicates results from polynomial approximation theory that support the use of *unbounded* kernels in the case of uniformly spaced, noise-free data. That said, we are not aware of any definitive analysis demonstrating the effect of these negative results on the NWRE or regularized strict interpolation RBFN in the regression context with which we are primarily concerned. Indeed, as we later present, the reasonable simulation results obtained using a Gaussian kernel in a range of practical applications suggest that the theoretical penalty imposed by conditions (24) and (25) in the Introduction need not be as grave as these studies may imply.

2. Both sets of conditions apply to the case of a strictly stationary random process $\{\boldsymbol{Z}(i)\}$ whose dependence structure is described by a *mixing* condition. For Eq. (1.8), $\{\boldsymbol{Z}(i)\}$ is assumed to follow a 2-α-*mixing* condition with geometric decay, that is,

$$\alpha^{(2)}(k) \triangleq \sup_{i \in \mathbb{Z}} \alpha(\sigma(\boldsymbol{Z}(i)), \sigma(\boldsymbol{Z}(i+k))) = \mathcal{O}(k^{\beta}) \qquad k \geq 1 \tag{1.26}$$

where, for two sub σ-fields $\mathcal{B}$, $\mathcal{C}$ of a common σ-field $\mathcal{A}$ with probability measure P, α is the strong mixing coefficient

$$\alpha = \alpha(\mathcal{B}, \mathcal{C}) \triangleq \sup_{B \in \mathcal{B}, C \in \mathcal{C}} |P(B \cap C) - P(B)P(C)| \tag{1.27}$$

For Eq. (1.6), $\{\boldsymbol{Z}(i)\}$ requires (not unexpectedly) the stronger *geometrically strong mixing (GSM)* condition, that is,

$$\alpha(k) \triangleq \sup_{j\in\mathbb{Z}} \alpha(\sigma(\{\boldsymbol{Z}(i)\}_{i=0}^{j}), \sigma(\{\boldsymbol{Z}(i)\}_{i=j+k}^{\infty})) = \mathcal{O}(\rho^k)$$

$$\text{for some } 0 \le \rho < 1 \quad (1.28)$$

Both of these mixing conditions are less restrictive than other types of mixing conditions commonly assumed, for example, ϕ and ρ mixing, and includes (trivially) the i.i.d. case.

3. For the convergence of Eq. (1.6), when K' is Lipschitz[5] and $p = p(z_1, \ldots, z_d) \in C_{2,d}(b)$ for some $b > 0$, where

$$C_{2,d}(b) \triangleq \left\{ f \in C_2(\mathbb{R}^d) \colon \|f\|_\infty < \infty \text{ and } \sup_{i,j=1,\ldots,d} \left\| \frac{\partial f}{\partial z_i \partial z_j} \right\|_\infty < b \right\} \quad (1.29)$$

$h_n = c_n(\log(n)/n)^{1/(d+4)}$, $c_n \overset{n\to\infty}{\to} c > 0$ assures that convergence for $D = D(n) \triangleq \{z \in \mathbb{R}^d : \|z\| \le n^\gamma\}\ \forall \gamma > 0$. This result implies a lower bound of $\mathcal{O}(n^{2d/(d+4)-\alpha}\ (c_n^d(\log n)^{d/(d+4)})^{-2})$ on the corresponding rate of convergence in Eq. (1.7). If K' is the Gaussian kernel and $\mathbb{E}[\|\boldsymbol{Z}(0)\|] < \infty$ (where $\{\boldsymbol{Z}(i)\} = \{\boldsymbol{Z}(i)\}_{i=0}^{\infty}$), then the convergence holds for $\mathbb{R}^d$ in place of D. If $c_n \searrow c$, the requirement on α for Eq. (1.7) to hold can be made independent of n by taking

$$\alpha > 1 + \max\left(1, \frac{d}{d+4}\left(1 - \frac{\log\log 2}{\log 2}\right) + \frac{\log[C/(c^d m)]}{\log 2}\right) \quad (1.30)$$

since

$$\frac{\log\left\{2C \Big/ \left[c^d\left(\frac{n}{\log n}\right)^{-d/(d+4)} m\right]\right\}}{\log n}$$

$$= \frac{\log[2C/(c^d m)]}{\log n} + \frac{\frac{d}{d+4}(\log n - \log\log n)}{\log n}$$

$$= \frac{\log[2C/(c^d m)]}{\log n} + \frac{d}{d+4}\left(1 - \frac{\log\log n}{\log n}\right)$$

$$\le \frac{\log 2 + \log[C/(c^d m)]}{\log 2} + \frac{d}{d+4}\left(1 - \frac{\log\log 2}{\log 2}\right)$$

$$\le 1 + \frac{\log[C/(c^d m)]}{\log 2} + \frac{d}{d+4}\left(1 - \frac{\log\log 2}{\log 2}\right)$$

[5]See note 5 in Appendix A.2.

4. For the convergence of Eq. (1.8), with some conditions on p and $\{\boldsymbol{Z}(i)\}$ [as described in the preamble to Theorem 2.1 of Bosq (1996)], $h_n = c_n n^{-1/(d+4)}$ where $c_n \overset{n\to\infty}{\to} c > 0$ is sufficient. Hence $nh_n^d = c_n^d n^{4/(d+4)}$, which implies that the rate of convergence in Eq. (1.12) is at least $\mathcal{O}(n^{1-\alpha})$ for $d \geq 1$. By a similar argument to that in the uniform case, if $c_n \searrow c$, the requirement on α for Eq. (1.12) to hold can be made independent of n by taking

$$\alpha > 1 + \max\left(v, \frac{d}{d+4} + \frac{\log[C/(c^d m)]}{\log 2}\right) \tag{1.31}$$

Some comments are now in order regarding the approximation theorems themselves:

1. The requirement that D, the region of approximation, be a compact subset of $\mathbb{R}^d$ such that p is strictly positive over D is common in both KDE and NWRE analysis (to ensure that the denominator is bounded away from zero). Given that approximation in regions of zero probability is of theoretical interest only, this requirement is not a practical impediment in applications. Later on, in Chapter 3, we shall extend the results to an increasing sequence of compact sets by imposing some extra conditions on the tail of the input process probability density.
2. The additional conditions required for the MS approximation case are quite mild; in particular, (1.9) is satisfied when the NWRE $\tilde{f}_n'$ and its MS approximating strict interpolation RBFN $\tilde{f}_{n,\infty}$ converge to continuous functions over D. Condition Eq. (1.10) is akin to the *local measure of dependence* κ_{ij}: $\mathbb{R}^d \times \mathbb{R}^d \to \mathbb{R}$ defined in Section 2.2 of Bosq (1996) as

$$\kappa_{ij}(\boldsymbol{x}, \boldsymbol{y}) \triangleq p_{ij}(\boldsymbol{x}, \boldsymbol{y}) - p(\boldsymbol{x})p(\boldsymbol{y}), \qquad \text{for } i \neq j \tag{1.32}$$

Now one possible sufficient condition in the preamble to Theorem 2.1 of Bosq (1996) is that (for all i and j) the κ_{ij} be Lipschitz when considered as a function over $\mathbb{R}^{2d}$. If this condition holds, then it can be shown that $\|\kappa_{ij}\|_\infty < M$, where M is a constant (dependent only upon the input dimension and the Lipschitz constant for κ_{ij}) times the first (and largest) 2-α-mixing coefficient $\alpha^{(2)}(1)$ of the input process $\{Z(i)\}$.[6] Because we assume that p is lower bounded over D by $m > 0$,

$$\sup_{x,y\in\mathbb{R}^d} |p_{ij}(x,y) - p(x)p(y)| < M \Rightarrow \sup_{x,y\in\mathbb{R}^d} \frac{p_{ij}(x,y)}{p(x)p(y)} < 1 + \frac{M}{m^2} \tag{1.33}$$

so Eq. (1.10) is fulfilled. The last condition, (1.11), requires that the minimum of the KDE $\tilde{p}_n$ over D go to zero no faster than some power of n. This condition is apparently necessary since, in general, the MS convergence Eq. (1.8) does not guarantee the corresponding pointwise convergence [so that an argument such as in Eq. (1.13) could be invoked], except in cases where $\tilde{p}_n$ and p are extremely regular (smooth), say members of a Sobolev space.[7]

3. In terms of the important constants governing the error bounds, we see that approximation over regions containing low probability events is more difficult than regions containing high probability events, demanding more training data. Ignoring the aiding factor of $(\log n)^{d/(d+4)}$ in the result of observation 3 above, n grows roughly at rate (no greater than) $m^{2d/(d+4)-\alpha}$ for a fixed level of error in the uniform approximation case. The corresponding rate in the MS approximation case, using the result of observation 4 above, is $m^{d/(d+4)-\alpha}$.
4. Since uniform approximation is a more stringent condition than MS approximation, it is not surprising that the asymptotic rate of convergence for the uniform approximation case is somewhat slower than that for the MS approximation case. For a given $\alpha > 2$ and assuming for simplicity $c_n \searrow c$, the rate of convergence in the uniform case is slower by a factor of $(n/\log n)^{2d/(d+4)}$ compared with the MS case. In both cases, however, the rates can be made arbitrarily fast by selecting α sufficiently large. This result is somewhat obvious from the construction of the approximating regularized strict interpolation RBFN, as the scaling factor n^α, which is applied to both the training output vector $\boldsymbol{y}$ and the input-dependent regularization parameter $\lambda_n(z)$, is designed to dominate (in norm) the regularization matrix $\boldsymbol{G}_n$. Eliminating $\boldsymbol{G}_n$ from Eq.

[6]This result is proven as Lemma 1.3 of (Bosq, 1996).
[7]See note 6 in Appendix A.2.

(15) in the Introduction gives weights that are consistent with those for the NWRE. Note that the minimum rate of output scaling is n^2 for uniform approximation versus n for MS approximation.

An immediate consequence of the uniform and MS approximation theorems is the trivial consistency (in the same modes) of regularized strict interpolation RBFNs in F_z when constructed to approximate known consistent NWREs. The argument could be made, however, that the family F_z with input-dependent regularization is not a natural one, especially in applications where, as will be described, a single regularization parameter determined once from a realized training set t_n is used over the entire domain of the regularized strict interpolation RBFN. While the introduction of F_z is motivated by its usefulness in the proofs, the arguments contained therein imply nonetheless that over any given compact set $D \subset \mathbb{R}^d$, the approximating regularized strict interpolation RBFNs have λ_n growing asymptotically at a rate *at least* $\Omega(n^{\alpha+4/(d+4)} \log^d n)$ for $\alpha > 2$ in the uniform case and $\Omega(n^{\alpha+4/(d+4)})$ for $\alpha > 1$ in the MS case, that is, at least roughly $\Omega(n)$ in both cases. By selecting the regularization parameter sequence to be asymptotically optimal, we can extend the MS consistency result from F_z to the ordinary noninput-dependent regularized strict interpolation RBFNs. To do so, the next section establishes a relationship between the MS fitting error $\tilde{R}_2$ and the usual global (MS) risk R_2.

1.3 RELATIONSHIP BETWEEN $\tilde{R}_2$ AND R_2

In this section, we show that under fairly mild conditions on $f, \tilde{f}_n$, and T_n, the MS fitting error $\tilde{R}_2$ converges to the true global risk R_2 for stochastic T_n when the input process $\{\boldsymbol{Z}(i)\}$ has a stationary measure $P = P_{\boldsymbol{Z}}$ and density p with respect to the Lebesgue measure. Note that the proof presented is quite general in that we do not assume any particular parametric form for either $\tilde{f}$ or the relationship between the input process $\{\boldsymbol{Z}(i)\}$ and the output process $\{Y(i)\}$. The proof is based on the following lemma for the convergence in probability of one nearest-neighbour distance. For simplicity, the lemma is demonstrated only for the case where $\boldsymbol{Z}$, the evaluation point, is independent of the training input sequence $\boldsymbol{Z}_n$, as defined below. The extension to the dependent case is fairly straightforward and will be developed in Chapter 3 for time series applications. Here the notation $X \sim P$ means that X is a RV defined with respect to the measure P.

Lemma 2. *Let* $\{\boldsymbol{Z}(1), \boldsymbol{Z}(2), \ldots, \boldsymbol{Z}(n), \boldsymbol{Z}\} \stackrel{\Delta}{=} \{\boldsymbol{Z}_n, \boldsymbol{Z}\} \sim P_{\boldsymbol{Z},\boldsymbol{Z}_n} \stackrel{\Delta}{=} P_{\boldsymbol{Z}_n} P_{\boldsymbol{Z}} = P_{\boldsymbol{Z}_n} P$, *that is,* $\boldsymbol{Z}$ *is independent of* $\boldsymbol{Z}_n$, *and* $P_{\boldsymbol{Z}(i)} = P$, $i = 1, 2, \ldots, n$, *that is,* $\boldsymbol{Z}_n$ *has stationary marginal measure P. Then for each* $\epsilon > 0$,

$$P_{\boldsymbol{Z},\boldsymbol{Z}_n}\left\{z, z_n: \min_{j=1,2,\ldots,n} \|z - z(j)\| > \epsilon\right\} \leq q^{n/2}(\epsilon) \stackrel{n\to\infty}{\to} 0 \tag{1.34}$$

where $\|\bullet\|$ *is the Euclidean norm in* $\mathbb{R}^d$ *and* $0 \leq q(\epsilon) < 1$ *is defined as*

$$q(\epsilon) \stackrel{\Delta}{=} \int_{\{(x,y)\,\in\,\mathbb{R}^d \times \mathbb{R}^d : \|x-y\| > \epsilon\}} dP(x)\, dP(y) \tag{1.35}$$

Proof. Let $\epsilon > 0$ be given. Set $A_{n,j}(\epsilon) \stackrel{\Delta}{=} \{z_{n+1}: \|z - z(j)\| > \epsilon\}$. We use the independence bound implied by the Cauchy–Schwarz inequality for the intersection of a finite collection of events $\{F_i\}_{i=1}^n$ defined with respect to a common probability measure P

$$P\left(\bigcap_{i=1}^n F_i\right) = \mathbb{E}\left[\prod_{i=1}^n I(F_i)\right] \leq \prod_{i=1}^n P^{1/2}(F_i) \tag{1.36}$$

where $I(\bullet)$ is the indicator function for event $\bullet$, so that

$$P_{\boldsymbol{Z}_{n+1}}\left(\bigcap_{j=1}^n A_{n,j}(\epsilon)\right) \leq \prod_{j=1}^n P^{1/2}_{\boldsymbol{Z},\boldsymbol{Z}(j)}(A_{n,j}) \tag{1.37}$$

From the stationarity of the marginal measure for $\{\boldsymbol{Z}(i)\}$ and the independence of $\boldsymbol{Z}$ from $\boldsymbol{Z}_n$, we have for $j = 1, 2, \ldots, n$

$$P_{\boldsymbol{Z},\boldsymbol{Z}(j)}(A_{n,j}) = \int_{\{(x,y)\,\in\,\mathbb{R}^d \times \mathbb{R}^d : \|x-y\| > \epsilon\}} dP(x) dP(y) \tag{1.38}$$

$$\stackrel{\Delta}{=} q(\epsilon) \tag{1.39}$$

It suffices therefore to show that for all $\epsilon > 0$, $q(\epsilon) < 1$. If, for a given ϵ, $q(\epsilon) = 1$, then $P_{\boldsymbol{Z},\boldsymbol{Z}(j)}(A^c_{n,j}(\epsilon)) = 0$. Since P (and hence $P_{\boldsymbol{Z},\boldsymbol{Z}(j)}$) is absolutely continuous with respect to the Lebesgue measure with density p, the support of p must have the nonzero Lebesgue measure. This fact combined with the a.e. continuity of p implies that an arbitrary open ball (of nonzero radius) centered about almost all points in the support of p must have nonzero measure with respect to P, hence $P_{\mathbf{Z},\mathbf{Z}(j)}(A^c_{n,j}(\epsilon)) = q(\epsilon) > 0$ for all $\epsilon > 0$. □

We can now verify claim A3 in the Introduction by showing the convergence of $\tilde{R}_2$ to R_2 [as defined in Eq. (4)]. Since $\tilde{R}_2$ appears to be a random-design quadrature formula for R_2, it may appear at first glance that a simple law of large numbers (LLN) type of argument would suffice. If, for instance, $\tilde{R}_2$ were evaluated over *another* set of sample input data $T'_m \stackrel{\Delta}{=} \{\mathbf{Z}'_i\}_{i=1}^m$ whose marginal conditional distribution $P_{T'_m|T_n} \stackrel{\Delta}{=} \prod_{i=1}^m P_{\mathbf{Z}'_i|T_n}$ satisfies $P_{\mathbf{Z}'_i|T_n} = P_{\mathbf{Z}|T_n}$ for $i = 1, 2, \ldots, m$, then conditioned on T_n, such LLNs would indeed apply to show that (in the notations of the Introduction) under appropriate conditions

$$\sup_{\tilde{f} \in \mathcal{F}} |\tilde{L}_2(\tilde{f}, f, T'_m) - R^*_2(\tilde{f}, f)| \stackrel{m \to \infty}{\longrightarrow} 0 \tag{1.40}$$

Since in our case T_n is used for both design and evaluation of the estimate $\tilde{f}_n$, this situation does not apply; instead we shall give a direct proof under some additional assumptions.

Theorem 2. *Assume that f is bounded as $|f| < L_f$ and Lipschitz with constant K_f over $\mathbb{R}^d$. If $\{\mathbf{Z}(i)\}$ and $\{Y(i)\}$ are such that:*

1. *There exists a positive constant L_p satisfying*

$$\sup_{z \in \mathbb{R}^d} |p(z)| < L_p \tag{1.41}$$

where p is the marginal density for $\{\mathbf{Z}(i)\}$.

2. *There exist positive constants L and K for the regularized strict interpolation RBFN estimate $\tilde{f}_n$ constructed from T_n satisfying for $n = 1, 2, \ldots,$*

$$\sup_{z \in \mathbb{R}^d} |\tilde{f}_n(z, T_n)| \leq L \; a.s.\text{–}P_{T_n} \tag{1.42}$$

$$K_n \leq K \; a.s.\text{–}P_{T_n} \tag{1.43}$$

where K_n is a Lipschitz constant for $\tilde{f}_n$.

then

$$|R_2(f, \tilde{f}_n) - \tilde{R}_2(f, \tilde{f}_n)| \overset{n\to\infty}{\to} 0 \ a.s.\text{-}P_{T_n} \tag{1.44}$$

Proof. For convenience of notation, let $v_{t_n}(\bullet) \triangleq (f(\bullet) - \tilde{f}_n(\bullet, t_n))^2$. Consider the ϵ cover $B_n(\epsilon)$ induced by a realized training sequence t_n, that is, $B_n(\epsilon) \triangleq \bigcup_{i=1}^{n} B_{n,i}(\epsilon)$, $B_{n,i}(\epsilon) \triangleq \{z, t_n: \|z - z(i)\| < \epsilon\}$. We may think of $B_n(\epsilon)$ as the set of all possible joint realizations of the training set and evaluation point such that the evaluation point is at least ϵ close to some training input point. An equivalent disjoint, that is, nonoverlapping, cover may be obtained by replacing $B_{n,i}(\epsilon)$ in the definition of $B_n(\epsilon)$ with $D_{n,i}(\epsilon) \triangleq B_{n,i}(\epsilon) \cap V_{z_n}(z(i))$ where $V_{z_n}(z(i))$ is the Voronoi cell[8] centered at $z(i)$ of the partition induced by the input sequence z_n contained in t_n. Decompose R_2 with respect to $B_n(\epsilon)$, where $\epsilon = \epsilon(n)$ (as will be explained later) so that

$$R_2(f, \tilde{f}_n) \triangleq \int_{z,t_n} v_{t_n}(z)dP(z, t_n)$$

$$= \int_{(z,t_n)\in B_n(\epsilon)} v_{t_n}(z)dP(z, t_n) + \int_{(z,t_n)\in B_n^c(\epsilon)} v_{t_n}(z)dP(z, t_n) \tag{1.45}$$

By the assumed boundedness of f and condition (1.42) on $\tilde{f}_n$, Lemma 2 implies that the latter integral can be made arbitrarily small for n sufficiently large since for any $\delta > 0$,

$$\int_{(z,t_n)\in B_n^c(\epsilon)} v_{t_n}(z)\, dP(z, t_n) \le L_v P_{Z,Z_n}(B_n^c(\epsilon)) \le \delta$$

$$\text{for} \quad n \ge 2\log\left(\frac{\delta}{L_v}\right) \Big/ \log q(\epsilon) \tag{1.46}$$

[8]See note 7 in Appendix A.2.

where $L_v = (L_f + L)^2$ is a global upper bound on v. For the former integral, we may write

$$\int_{(z,t_n)\in B_n(\epsilon)} v_{t_n}(z)dP(z,t_n) = \sum_{i=1}^{n} \int_{(z,t_n)\in D_{n,i}(\epsilon)} v_{t_n}(z)dP(z,t_n)$$

$$\overset{\le}{\underset{+}{\overset{-}{\ge}}} \sum_{i=1}^{n} \int_{(z,t_n)\in D_{n,i}(\epsilon)} (v_{t_n}(z(i)) \mp K_v\epsilon)\, dP(z,t_n)$$

$$\overset{\le}{\underset{+}{\overset{-}{\ge}}} \int_{t_n} \left[\sum_{i=1}^{n} (v_{t_n}(z(i)) \mp K_v\epsilon) P_{Z|T_n}(D_{n,i}(\epsilon)|T_n = t_n) \right] dP(t_n) \qquad (1.47)$$

by the definition of $D_{n,i}(\epsilon)$ and the fact that f and $\tilde{f}_n$ are bounded and Lipschitz implies the same for v with Lipschitz constant not greater than $K_v \triangleq 2(L_f + L)(K_f + K)$. Here the notation $a \overset{\le}{\underset{+}{\overset{-}{\ge}}} f(b \mp c)$ is shorthand for the double inequality $f(b-c) \le a \le f(b+c)$, where f is an expression containing $b \mp c$. The remainder term containing $K_v\epsilon$ can be bounded uniformly over all possible training realizations t_n (equivalently, over all possible training input realizations z_n) since

$$P_{Z|T_n}(D_{n,i}(\epsilon)|T_n = t_n) \le L_p(2\epsilon)^d \qquad \forall t_n \qquad (1.48)$$

by Eq. (1.41) and where we have used the (Euclidean) volume of a d-dimensional cube in $\mathbb{R}^d$ with edge 2ϵ to upper bound the volume of the corresponding closed ball of radius ϵ. Hence we have the simultaneous lower and upper bounds

$$\left| \int_{z,t_n\in D_n(\epsilon)} v_{t_n}(z)dP(z,t_n) - \int_{t_n} \sum_{i=1}^{n} v_{t_n}(z(i)) P_{Z|T_n}(D_{n,i}(\epsilon)|T_n = t_n) dP(t_n) \right|$$
$$\le nK_vL_p\epsilon(2\epsilon)^d \qquad (1.49)$$

For the remainder term $r(n) \triangleq nK_vL_p\epsilon(2\epsilon)^d$ to vanish as $n \to \infty$, we require that $\epsilon^{d+1} = \mathcal{O}(1/n^{1+\beta})$ for some $\beta > 0$. At the same time, the inequality

$$q(\epsilon) = 1 - P_{Z,Z(i)}(B_{n,i}(\epsilon)) \ge 1 - L_p(2\epsilon)^d \qquad (1.50)$$

implies that we cannot let ϵ decrease too quickly as $n \to \infty$ if Eq. (1.46) is to be satisfiable for $\delta = \delta(n) = \mathcal{O}(1/n^{\alpha})$ with $\alpha > 0$, since for x small,

$\log(1-x) \underset{n\to\infty}{\approx} -x$. In other words, for Eq. (1.46) to hold with $L_v > \delta(n) \xrightarrow{n\to\infty} 0$, it is necessary that $\epsilon^d = \Omega(1/n^{1-\gamma})$ for some $\gamma \in (0,1)$. Equating the two exponents gives the relationship between β and γ as

$$0 < \beta < \frac{1}{d} \qquad \gamma = \frac{1-\beta d}{1+d} \tag{1.51}$$

Returning to the integral term in Eq. (1.49), we recombine the iterated expectation and note that

$$\left| \int_{t_n} \sum_{i=1}^{n} v_{t_n}(z(i))dP(t_n) - \int_{t_n} \sum_{i=1}^{n} v_{t_n}(z(i))P_{Z|T_n}(D_{n,i}(\epsilon)|T_n = t_n)dP(t_n) \right|$$

$$\leq \sup_{i=1,2,\ldots,n} |v_{t_n}(z(i))| \left| \sum_{i=1}^{n} \left(\frac{1}{n} - \int_{z,t_n \in D_{n,i}(\epsilon)} dP(z,t_n) \right) \right| \tag{1.52}$$

$$\leq L_v \left| 1 - \sum_{i=1}^{n} P_{Z,T_n}(D_{n,i}(\epsilon)) \right|$$

$$\leq L_v |1 - P_{Z,T_n}(B_n(\epsilon))| = L_v q^{n/2}(\epsilon)$$

$$\leq \delta \tag{1.53}$$

where we have again invoked Lemma 2 in the last line for δ as defined in Eq. (1.46). Combining the inequalities (1.46), (1.49), and (1.53) yields

$$\begin{aligned} |R_2(f,\tilde{f}_n) - \tilde{R}_2(f,\tilde{f}_n)| &\leq r(n) + 2\delta(n) \\ &= \mathcal{O}(n^{-\beta}) + \mathcal{O}(n^{-\alpha}), \qquad 0 < \beta < 1/d \quad \alpha + \beta < 1 \end{aligned} \tag{1.54}$$

where the condition $\alpha + \beta < 1$ is required for Eq. (1.46) to hold. This result implies that the asymptotic rate of convergence of $\tilde{R}_2(f,\tilde{f}_n)$ to $R_2(f,\tilde{f}_n)$ can be made arbitrarily close to (but strictly less than) $\mathcal{O}(n^{-1/d})$, from which the desired conclusion follows. □

As an aside, the conditions required for the theorem to hold are not quite as restrictive as they may appear at first sight; in particular, conditions (1.42) and (1.43) hold, for example, when the estimate $\tilde{f}_n$ converges uniformly a.s.–P_{T_n} to a bounded, Lipschitz function. Although, as mentioned in the Introduction, Plutowski et al. (1994) show a similar result in their proof that the (ordinary) cross-validation [(O)CV] proxy for the global MSE R_2 converges (unbiasedly) a.s., the proof here relaxes the requirement for i.i.d. samples to allow dependent samples from a bounded and stationary input process density. We also provide an estimate of the rate of convergence as approximately $\mathcal{O}(n^{-1/d})$, which illustrates the "curse of dimensionality" prevalent in high-dimensional estimation problems [a similar inverse dependence of the sample exponent on the input dimensionality can be seen in the KDE/NWRE results of Bosq (1996) cited earlier]. While at this point we are not aware of any other rigorous demonstrations of the convergence of the MS fitting error $\tilde{R}_2$ to the global MSE R_2 under our chosen circumstances, existing strong law of large numbers (SLLN) results appear to support a possible improvement to an exponential rate of convergence independent of (or only weakly dependent on) the input process dimensionality.

1.4 IMPLICATIONS FOR REGULARIZED STRICT INTERPOLATION RBFN ESTIMATION

1.4.1 Asymptotic Optimality with Respect to Mean-Squared Error

The convergence of $\tilde{R}_2$ to R_2 demonstrated above is desirable because the asymptotic optimality of the CV and generalized cross-validation (GCV) parameter selection methods with respect to the loss $\tilde{L}_2$ and the risk $\tilde{R}_2$ for linear estimates has been studied under various scenarios. The ones mentioned below follow the usual practice of assuming that one is observing a sequence of fixed means that have been additively corrupted by an independent sequence of bounded variance, zero-mean errors. At first glance, this model may appear somewhat restrictive, but it can include the stochastic T_n case described in the Introduction by setting the error $\epsilon(i) \stackrel{\Delta}{=} Y(i) - f(\boldsymbol{Z}(i))$, where f is the (assumed stationary) regression function $f(\bullet) \stackrel{\Delta}{=} \mathbb{E}[Y(i) \mid X(i) = \bullet]$, and conditioning on a fixed input sequence z_n, that is, the problem is reduced to the case of noisy T_n. Note that after this change, stochastic T_n with i.i.d. samples $(\boldsymbol{Z}(i), Y(i))$ lead to a conditionally heteroskedastic model, that is, one in which the errors $\epsilon(i)$ are independent zero mean but not identically distributed, as their variances $\sigma^2(i)$ are conditional on the corresponding input realization $z(i)$ [for example, see the discussion in Section 4.10 of Eubank (1998)]. If this dependence on $z(i)$ is known, however, as

$$\sigma^2(i) = \frac{\sigma^2}{\psi^2(z(i))} \tag{1.55}$$

where ψ is a known function, then replacing each pair $(z(i), y(i))$ in t_n with the pseudo-data $(z(i), \psi(z(i))y(i))$ gives an equivalent homoskedastic error with respect to $\psi \cdot f$, that is, $\epsilon'(i) \stackrel{\Delta}{=} \psi(\boldsymbol{Z}(i))Y(i) - \psi(\boldsymbol{Z}(i))f(\boldsymbol{Z}(i))$ is i.i.d. zero mean with common variance σ^2 [for example, see Section 2.1 of Gallant (1987)]. While this result is theoretically appealing, its practice when ψ is unknown effectively adds the complication of estimating the conditional variance of $Y(i)$ given $\boldsymbol{Z}(i) = z(i)$, which may only be feasible in the case that the heteroskedasticity is mild, that is, ψ is a relatively smooth function. In any case, the scenarios are:

1. Homoskedastic errors and continuous parameter set (Li, 1985).
2. Homoskedastic errors and discrete parameter set (with cardinality possibly increasing in the number of data, subject to certain conditions) (Li, 1987).
3. Heteroskedastic errors and discrete parameter set (with cardinality possibly increasing in the number of data, subject to certain conditions) (Andrews, 1991).

While CV has been shown to be generally asymptotically optimal under all of the above scenarios, the same is known to be true for GCV only in the first two scenarios; more specifically, the third study gave sufficient conditions for asymptotic optimality, which are not satisfied by GCV for ridge regression[9] problems such as those involving the selection of the regularization parameter for strict interpolation RBFNs. This current drawback of GCV can be theoretically addressed using one of the variance-equalizing methods as described above; in practice, however, the actual loss of performance due to heteroskedasticity with GCV varies and is not inevitably significant (see the discussion in item 3 of Section 2.5).

For our purposes, if we assume that the estimates $\tilde{f}_n$ satisfy the uniform boundedness condition Eq. (1.42) of Theorem 2, then the input sequence conditioning of the standard results can be avoided. Let us define the input sequence conditioned version of $\tilde{R}_2$ as

$$\tilde{R}_2(\lambda; z_n) \stackrel{\Delta}{=} \mathbb{E}_{T_n|Z_n}\left[\frac{1}{n}\sum_{i=1}^{n} v_{T_n}(\boldsymbol{Z}(i)) \middle| \boldsymbol{Z}_n = z_n\right]$$

$$\text{so that} \quad \tilde{R}_2(\lambda) = \mathbb{E}_{Z_n}[\tilde{R}_2(\lambda; \boldsymbol{Z}_n)] \quad (1.56)$$

where v_{T_n} is as defined in the proof of Theorem 2 with $\tilde{f}_n$ designed using the regularization parameter λ. On occasion, we shall write $\tilde{R}_2$ as $\tilde{R}_{2,n}$ to

[9]See note 8 in Appendix A.2.

emphasize its dependence on n when it is not clear from context, for example, when discussing $\tilde{R}_2$ for fixed λ. Now the $\tilde{R}_2$ a.o. of a regularization parameter sequence $\{\tilde{\lambda}_n\}$ means that, given any infinite-length input sequence realization z_∞ with subsequence $z_n = \{z_\infty\}_{1:n}$,

$$\frac{\tilde{R}_{2,n}(\tilde{\lambda}_n; z_n)}{\inf_{\lambda \in \mathbb{R}^+} \tilde{R}_{2,n}(\lambda; z_n)} \overset{n\to\infty}{\longrightarrow} 1 \tag{1.57}$$

Because the convergence occurs for all possible realizations of z_∞ and hence z_n, it holds i.p.–P_{Z_n}, that is, given any $\epsilon > 0$,

$$P_{Z_n}\left\{z_n: \left|\frac{\tilde{R}_{2,n}(\tilde{\lambda}_n; z_n) - \inf_{\lambda \in \mathbb{R}^+} \tilde{R}_{2,n}(\lambda; z_n)}{\inf_{\lambda \in \mathbb{R}^+} \tilde{R}_{2,n}(\lambda; z_n)}\right| > \epsilon\right\} \overset{n\to\infty}{\to} 0 \tag{1.58}$$

By the uniform boundedness of $\tilde{f}_n$, the denominator of the above event must be upper bounded by some $L > 0$ for all n so that Eq. (1.58) implies

$$P_{Z_n}\left\{z_n: \left|\tilde{R}_{2,n}(\tilde{\lambda}_n; z_n) - \inf_{\lambda \in \mathbb{R}^+} \tilde{R}_{2,n}(\lambda; z_n)\right| > L\epsilon\right\} \overset{n\to\infty}{\to} 0 \tag{1.59}$$

that is, $\tilde{R}_{2,n}(\tilde{\lambda}_n; \boldsymbol{Z}_n)$ converges to $\inf_{\lambda \in \mathbb{R}^+} \tilde{R}_{2,n}(\lambda; \boldsymbol{Z}_n)$ i.p.–P_{Z_n}. Then, using the fact that for uniformly integrable $\tilde{f}_n$, convergence i.p.–P_{Z_n} also implies convergence in $L_1(P_{Z_n})$ [for example, see Lemma 4.6.6 of Gray (1987)], we may conclude that

$$0 \leq \tilde{R}_2(\tilde{\lambda}_n) - \inf_{\lambda \in \mathbb{R}^+} \tilde{R}_2(\lambda) \overset{n\to\infty}{\to} 0 \tag{1.60}$$

As an aside, note that the convergence specified by asymptotic optimality and Eq. (1.60), while similar, are not identical. In the former case, the *rate* of convergence is at least that of the denominator $\inf_{\lambda \in \mathbb{R}^+} \tilde{R}_{2,n}(\lambda; z_n)$ to zero, assuming that the class of estimates containing $\tilde{f}_n$ is $\tilde{R}_2$ consistent [that is, given any z_∞, $\inf_{\lambda \in \mathbb{R}^+} \tilde{R}_{2,n}(\lambda; z_n) \overset{\text{i.p. } n\to\infty}{\to} 0$]; in the latter case, the rate of convergence is not known.

Using the unconditional asymptotic optimality of Eq. (1.60), it is now possible to show that an asymptotically optimal regularization parameter sequence $\{\lambda_n\}$ also asymptotically minimizes the global MSE R_2, that is, minimum mean-square error (MMSE) estimation over the class of regularized strict interpolation RBFNs is asymptotically realized.

Corollary 1. *Let the strict interpolation RBFN regularization parameter sequence* $\{\tilde{\lambda}_n\}$ *be selected by a procedure under conditions for which the sequence is asymptotically optimal in the MS fitting error (risk)* $\tilde{R}_2 = \tilde{R}_{2,n}$ *as per Eq. (1.60). In addition, assume that the conditions for Theorem 2 hold. Then*

$$\left| R_2(\tilde{\lambda}_n) - \inf_{\lambda \in \mathbb{R}^+} R_2(\lambda) \right| \stackrel{n\to\infty}{\rightarrow} 0 \tag{1.61}$$

that is, $\{\tilde{\lambda}_n\}$ *is a consistent sequence of estimates for the global MSE-minimizing regularization parameter.*

Proof. By the triangle inequality, we have

$$\begin{aligned} \left| R_2(\tilde{\lambda}_n) - \inf_{\lambda \in \mathbb{R}^+} R_2(\lambda) \right| &\leq |R_2(\tilde{\lambda}_n) - \tilde{R}_{2,n}(\tilde{\lambda}_n)| \\ &\quad + \left| \tilde{R}_{2,n}(\tilde{\lambda}_n) - \inf_{\lambda \in \mathbb{R}^+} \tilde{R}_{2,n}(\lambda) \right| \\ &\quad + \left| \inf_{\lambda \in \mathbb{R}^+} \tilde{R}_{2,n}(\lambda) - \inf_{\lambda \in \mathbb{R}^+} R_2(\lambda) \right| \end{aligned} \tag{1.62}$$

For the last term, we may use the bound

$$\left| \inf_{\lambda \in \mathbb{R}^+} \tilde{R}_{2,n}(\lambda) - \inf_{\lambda \in \mathbb{R}^+} R_2(\lambda) \right| \leq \sup_{\lambda \in \mathbb{R}^+} |\tilde{R}_{2,n}(\lambda) - R_2(\lambda)| \tag{1.63}$$

[since in the case that the infima occur for different λ, say $\tilde{\lambda}^*$ for $\tilde{R}_{2,n}$ and λ^* for R_2, one of $|\tilde{R}_{2,n}(\tilde{\lambda}^*) - R_2(\tilde{\lambda}^*)|$ or $|\tilde{R}_{2,n}(\lambda^*) - R_2(\lambda^*)|$ must be at least as large as the left-hand side term being bounded]. Then, using this bound, the first and last terms on the right-hand side of Eq. (1.62) converge to zero a.s.-P_{T_n} by virtue of Eq. (1.44) (which holds independent of the choice of λ for the last term), while the middle term does the same by Eq. (1.60), thus completing the proof. □

1.4.2 Consistency with Respect to Mean-Squared Error over Compacta

By the previous theorems, a heuristic argument for the MS consistency of regularized strict interpolation RBFNs with $\tilde{\lambda}_n$ selected by a suitable asymptotically optimal parameter estimation procedure can be posited as follows: suppose that conditions are such that a NWRE f_n' is MS consistent, that is, $R_2(f, \tilde{f}_n') \overset{n\to\infty}{\to} 0$ in some mode m. Then $\tilde{f}_{n,\infty}$, the a.s. uniform strict interpolation RBFN approximation of $\tilde{f}_n'$ constructed according to Theorem 1, must also be MS consistent, that is, $R_2(f, \tilde{f}_{n,\infty}) \overset{n\to\infty}{\to} 0$ (in mode m). The a.s. convergence of $\tilde{R}_2$ to R_2 further implies that the MS fitting error $\tilde{R}_2(f, \tilde{f}_{n,\infty}) \overset{n\to\infty}{\to} 0$ (in the same mode m). At the same time, the MS fitting error $\tilde{R}_2(f, \tilde{f}_n)$ of the strict interpolation RBFN $\tilde{f}_n$ that is the same strict interpolation RBFN as $\tilde{f}_{n,\infty}$ except with $\tilde{\lambda}_n$ chosen asymptotically optimally must (by the definition of asymptotic optimality) be asymptotically no greater than $\tilde{R}_2(f, \tilde{f}_{n,\infty})$. Hence $\tilde{R}_2(f, \tilde{f}_n) \overset{n\to\infty}{\to} 0$ (in mode m). Once again, the a.s. convergence of $\tilde{R}_2$ to R_2 can be invoked to conclude that $R_2(f, \tilde{f}_n) \overset{n\to\infty}{\to} 0$ (in mode m), that is, regularized strict interpolation RBFNs with λ_n selected asymptotically optimally are MS consistent. Before giving a more detailed proof, it would be convenient to extend the results in Theorem 1 to hold globally in some sense rather than locally over a fixed compact set. Although this modification can be effected for both the uniform and MS strict interpolation RBFN approximation cases, it turns out that the simpler extension in the uniform case is sufficient for our intended applications. At the same time, we shall also (somewhat trivially) allow a general exponent $\theta \geq 1$. We begin by modifying the definition found in Section 3.2.2 of Bosq (1996) to say that a sequence $\{S_n\}$ of compact sets in $\mathbb{R}^d$ is *regular* (with respect to a density p) if there exists a sequence $\{\beta_n\}$ and a monotonically increasing sequence $\{\rho_n\}$ of strictly positive real numbers such that for each n,

$$\inf_{z \in S_n} p(z) \geq \beta_n \quad \text{and} \quad \operatorname{diam}(S_n) \triangleq \sup_{x,y \in S_n} \|x - y\| \leq \rho_n \tag{1.64}$$

[in Bosq (1996), ρ_n is set to n^{γ}; the reason for our generalization will become clear shortly]. A probability density p for a measure P is said to be *regular in probability* if it admits a regular sequence $\{S_n\}$ with $\lim_{n\to\infty} P(S_n) = 1$. We shall also define a tail condition for a density p as

$$\exists R > 0 \quad \text{such that} \quad p(z) \geq \mu(\|z\|) \text{ when } \|z\| > R \tag{1.65}$$

where $\mu\colon \mathbb{R}^+ \to \mathbb{R}^+$ is a bicontinuous[10] monotonically decreasing function. We can now give the following

Theorem 3. *Using the same conditions and notations as for Eq. (1.7), if p is regular in probability with regular sequence $\{S_n\}$ such that Eq. (1.64)*

[10]A function f is *bicontinuous if f* is one-to-one and continuous so that it has a continuous inverse.

and Eq. (1.65) jointly satisfy

$$\lim_{n\to\infty} \frac{\log(1/\mu(\rho_n))}{\log n} < \infty \tag{1.66}$$

then a regularized strict interpolation RBFN $\tilde{f}_{n,\infty} \in F_z$ *may be constructed such that*

$$\sup_{z\in S_n} |\tilde{f}_{n,\infty}(z) - \tilde{f}'_n(z)|^{\theta} \overset{n\to\infty}{\rightarrow} 0 \; a.s.-P_{T_n} \tag{1.67}$$

for $\theta \in [1, \infty)$, *that is, the uniform approximation holds globally i.p.–P for almost all training sets* T_n.

Proof. For $\theta > 1$, the bound of Eq. (1.7) still holds after raising the order terms to the power θ. Let $\{S_n\}$ be a regular sequence for p with sequences $\{\beta_n\}$ and $\{\rho_n\}$ satisfying Eq. (1.64) and μ and R as defined in Eq. (1.65). In the proof of Eq. (1.7), we replace m, the lower bound on the density p over D, with β_n which, by the previous conditions, satisfies

$$\beta_n \geq \mu(\rho_n), \qquad n > N_1 \text{ where } N_1 \text{ is such that } \rho_n > R \quad \text{for all } n > N_1 \tag{1.68}$$

The required condition on α for the approximation error to decrease asymptotically to zero then becomes

$$\alpha > \max\left(2, \frac{\log(2C/h_n^{d}) + \log(1/\mu(\rho_n))}{\log n}\right), \qquad n > N_1 \tag{1.69}$$

which can be made independent of n as in (1.30) because of the relative stability condition (1.66). Since p is regular in probability, the uniform approximation holds with arbitrarily high $\boldsymbol{Z}$ probability when Eq. (1.69) is fulfilled, completing the proof. □

Note that the joint condition Eq. (1.66) is equivalent to $\mu(\rho_n) = \Omega(n^{-q})$ for some $q > 0$. Examples of densities regular in probability that satisfy Eq. (1.66) include Gaussians and their mixtures [with $\rho_n = r\sqrt{\log n}$ and $\mu(\bullet) = \exp(-k\|\bullet\|^2)$ for some strictly positive reals r and k] and compactly supported, bounded, continuous densities with polynomial tails, for example, uniform density.

With the above tools in place, we may derive an MS consistency result for regularized strict interpolation RBFNs with asymptotically optimal regularization sequence $\{\tilde{\lambda}_n\}$. The notion of "corresponding NWRE" was defined previously in Section 1.2. For greater generality, we do not rely on the assumption that $\boldsymbol{Z}$ is independent of the training set T_n in this proof, although, as we shall discuss in Chapter 3, the same can be achieved for the previous proofs with only slightly more effort. The following lemma concerning a triangle inequality-type bound for R_2 will be useful:

Lemma 3. *For f, g, and $h \in L_2(\mathbb{R}^d, P_{\boldsymbol{Z},T_n})$,*

$$|R_2(f,h) - R_2(g,h)| \leq R_2(f,g) + 2\sqrt{R_2(f,g)R_2(g,h)} \qquad (1.70)$$

Proof. As a shorthand, let $\|\bullet\|_P \triangleq \mathbb{E}_{\boldsymbol{Z},T_n}[\bullet]$. Then

$$\begin{aligned} |R_2(f,h) - R_2(g,h)| &= |\|(f-h)^2\|_P - \|(g-h)^2\|_P| \\ &\leq \||f-h+g-h| \cdot |f-g|\|_P \\ &\leq \|(|f-g| + 2|g-h|) \cdot |f-g|\|_P \\ &\leq \|(f-g)^2\|_P + 2\|(f-g)^2\|_P^{1/2}\|(g-h)^2\|_P^{1/2} \end{aligned} \qquad (1.71)$$

which is the desired conclusion. □

Theorem 4. *Assume that Theorem 2 holds and let the stationary marginal density p for $\{\boldsymbol{Z}(i)\}$ be regular in probability. Then the regularized strict interpolation RBFN with R_2 a.o. regularization parameter sequence $\{\tilde{\lambda}_n\}$ [per (1.60)] is globally MS consistent whenever the corresponding NWRE is globally MS consistent.*

Proof. Let $\tilde{f}'_n$ be the globally MS consistent NWRE and $\tilde{f}_{n,\infty}$ the corresponding uniform approximation to $\tilde{f}'_n$ constructed by Theorem 1 to satisfy Eq. (1.7) and its extended version Eq. (1.67) in Theorem 3. Applying the preceding lemma,

$$|R_2(\tilde{f}_{n,\infty}, f) - R_2(\tilde{f}'_n, f)| \leq R_2(\tilde{f}_{n,\infty}, \tilde{f}'_n) + 2\sqrt{R_2(\tilde{f}'_n, f)R_2(\tilde{f}_{n,\infty}, \tilde{f}'_n)} \qquad (1.72)$$

Because $\tilde{f}'_n$ is globally MS consistent, the right-hand side above vanishes so long as $R_2(\tilde{f}_{n,\infty}, \tilde{f}'_n)$ does too. Choosing $\theta = 2$ in Theorem 3, we have the convergence

$$|\tilde{f}_{n,\infty}(\boldsymbol{Z}) - \tilde{f}'_n(\boldsymbol{Z})|^2 \overset{n\to\infty}{\to} 0 \quad \text{i.p.}-\text{P}_{\boldsymbol{Z},T_n} \tag{1.73}$$

which, by the assumption that the $\tilde{f}_{n,\infty}$ and $\tilde{f}'_n$ are uniformly integrable, also implies the corresponding MS convergence

$$R_2(\tilde{f}_{n,\infty}, \tilde{f}'_n) \overset{n\to\infty}{\to} 0 \tag{1.74}$$

Hence,

$$|R_2(\tilde{f}_{n,\infty}, f) - R_2(\tilde{f}'_n, f)| \overset{n\to\infty}{\to} 0 \Rightarrow R_2(\tilde{f}_{n,\infty}, f) \overset{n\to\infty}{\to} 0 \tag{1.75}$$

so that $\tilde{f}_{n,\infty}$ is also globally MS consistent. From this and the a.s. convergence of $\tilde{R}_2$ to R_2, it follows that

$$\tilde{R}_2(\tilde{f}_{n,\infty}, f) \overset{n\to\infty}{\to} 0 \tag{1.76}$$

On the other hand, by definition, for each n,

$$\tilde{R}_2(\tilde{f}_{n,\infty}, f) \geq \inf_{\lambda \in \mathbb{R}^+} \tilde{R}_{2,n}(\lambda) \Rightarrow \inf_{\lambda \in \mathbb{R}^+} \tilde{R}_{2,n}(\lambda) \overset{n\to\infty}{\to} 0 \tag{1.77}$$

and, reversing the previous reasoning, by the definition of asymptotic optimality in Eq. (1.60) and the a.s. convergence of $\tilde{R}_2$ to R_2,

$$\tilde{R}_2(\tilde{f}_n, f) \overset{n\to\infty}{\to} 0 \Rightarrow R_2(\tilde{f}_{n,\infty}, f) \overset{n\to\infty}{\to} 0 \tag{1.78}$$

□

1.5 SUMMARY AND DISCUSSION

We have shown how strict interpolation RBFNs can be made to behave asymptotically as NWREs by a special choice of input-dependent regularization parameter sequence, leading to a limited consistency result for such constrained strict interpolation RBFNs. To deduce the corresponding consistency for the usual class of strict interpolation RBFNs where the regularization parameter sequence is not input-dependent but rather chosen on the basis of the available training data, we introduced the results for asymptotically optimal regularization parameter sequences. After we proved the convergence of the MS fitting error, the criterion for asymptotic optimality, to the true global risk or MSE in which we are interested, these results were applied to demonstrate how the consistency of the constrained strict interpolation RBFNs implies the same for regularized strict interpolation RBFNs designed in the usual way with asymptotically optimal regularization parameter sequence. Thus in Theorem 4, we have verified claim A4 made in the Introduction. More precisely, we have shown that regularized strict interpolation RBFNs with asymptotically optimal regularization parameter sequence are (globally) MS consistent under conditions for which the NWRE is both MS consistent and uniformly approximable according to Theorem 3. We have therefore at least partially answered the question of a justifiable rather than *ad hoc* design procedure for RBFNs. Following are some other relevant remarks in this regard:

1. The reasoning supporting MS consistency in the regularized strict interpolation RBFN case carries over to the regularized *random centers* RBFN case [Eq. (19) in the Introduction] if the center (equivalently, basis function) selection method used is proven to be R_2-consistent. More precisely, denote the center selection policy by n and the number of centers it selects from the N available training data in T_N as $n(N) \leq N$. Then it would be sufficient that

$$\inf_{\lambda \in \mathbb{R}^+} R_{2,n(N)}(\lambda) \overset{N \to \infty}{\to} 0 \qquad (1.79)$$

   for a regularized random centres RBFN with centers chosen by policy n and asymptotically optimal regularization parameter sequence $\{\tilde{\lambda}_n\}$ to be MS consistent. The burden of this additional proof is avoided in the strict interpolation, that is, full centers, RBFN construction.

2. In addition to being $\tilde{R}_2$–a.o., CV parameter selection procedures are also known to be $\tilde{L}_2$–a.o. i.p.–P_{T_n}, that is,

$$\frac{\tilde{L}_2(\tilde{\lambda}_n; z_n)}{\inf_{\lambda \in \mathbb{R}^+} \tilde{L}_2(\lambda; z_n)} \overset{n \to \infty}{\to} 1 \text{ i.p.}-P_{T_n} \qquad (1.80)$$

   so that the true loss for a *particular* training input sequence z_n is correctly estimated. This optimality condition is stronger than

$\tilde{R}_2$–a.o. and can be useful when the inputs in z_n represent especially important evaluation points, for example, the modes of the probability measure $P_{\mathbf{Z}}$ in the stationary input case. In the proof of Theorem 4, however, this $\tilde{L}_2$–a.o. is not necessary.

3. An appropriate level of smoothing through the choice of regularization parameter sequence is central to the consistency of the regularized strict interpolation RBFN. In fact, Theorem 8 of Golitschek and Schumaker (1990) effectively implies item A5 from the Introduction. That is, for noisy T_n in scenario 1 of Section 1.4.1, RBFNs designed with positive definite kernel function and $\lambda_n = 0$ for all n cannot be MS consistent unless the number of basis functions grows more slowly than the number of data. For any training realization t_n, applying the said theorem yields

$$\tilde{R}_2(0; z_n) = \mathbb{E}[\epsilon^2(i)] \stackrel{\Delta}{=} \sigma^2, \quad \forall z_n \tag{1.81}$$

which is obvious since for any strict interpolation RBFN design with $\lambda = 0$, the predicted output vector $\tilde{\boldsymbol{y}}_n$ equals the training output vector $\boldsymbol{y}_n$. By the convergence of $\tilde{R}_2$ to R_2, this result implies that such unregularized networks must have MSE bounded away from zero for n sufficiently large, hence claim A5 of the Introduction is verified. This result merely confirms the artificial neural network (ANN) intuition that exact fitting of noisy t_n with n basis functions leads to an overfit and hence poor generalization. On the other hand, the situation for $\lambda_n = \mathcal{O}(n^\alpha)$ for some $\alpha > 2$ as in Theorem 3 is not quite as clear-cut, since the NWREs being approximated are known to be (MS) consistent. What we can conclude, however, is that such an effective choice of asymptotically increasing regularization parameter sequence (as for the NWRE) is not optimal in terms of minimizing either the MS fitting error $\tilde{R}_2$ or the actual loss $\tilde{L}_2$. More definitive statements require a specific analysis of the eigenvalues of the interpolation matrix $\boldsymbol{G}_n$ (Golitschek and Schumaker, 1990). Section 4.4 of Wahba (1990) performs this for the smoothing spline case and argues that $\tilde{R}_2$ cannot vanish unless, among other conditions, $\lambda_n \xrightarrow{n \to \infty} 0$.

2

PROBABILITY ESTIMATION AND PATTERN CLASSIFICATION

2.1 INTRODUCTION

In this chapter, we prove certain results pertaining to the strict interpolation RBFN estimation of posterior probabilities using least-squares regression of indicator functions and the subsequent implications for pattern classification. Since these results are relatively simple to establish using the theoretical tools developed in Chapter 1, this material is included primarily for completeness and continuity in demonstrating the generality of the tools. The treatment is therefore somewhat briefer and more theoretically oriented than those that follow in Chapter 3. Furthermore, we should add the caveat that the results obtained do not justify *a priori* such least-squares-based techniques as the most suitable or natural ones for classification problems; indeed, if a maximum likelihood estimate of the posterior probabilities is desired, then *logistic*, not least-squares, regression is appropriate (McLachan, 1992). It is also intuitively clear that a posterior probability estimate that is optimal with respect to a squared-error criterion such as R_2 need not be optimal with respect to the average probability of classification error when used in an approximate

Bayes decision rule, as described below.[1] That said, since least-squares procedures are widely used in practice due to their analytical as well as computational tractability, a clearer understanding of their theoretical properties in such applications is arguably useful.

2.2 PROBLEM DESCRIPTION

In this section, we set the notations and the basic framework within which we shall consider the related problems of probability estimation and pattern classification. The type of probability estimates that we shall be considering are the so-called *posterior* or conditional probabilities $P_A(\boldsymbol{x}) \triangleq \Pr(A|\boldsymbol{X} = \boldsymbol{x})$, where $\boldsymbol{X} = \boldsymbol{X}(\omega)$ is an $\mathbb{R}^d$-valued RV generically known as the *feature* and A is an event, both defined with respect to a common underlying sample space Ω. That such posterior probabilities arise naturally in pattern classification stems from the well-known *Bayes rule* for *optimal*, that is, minimum average error, decisions when Ω is a discrete sample space representing the occurrence of each class. Without loss of generality, if there are M classes, we can equivalently enumerate the classes via a discrete RV $Y \in C = \{1, 2, \ldots, M\}$. Then given a realized feature $\boldsymbol{x}$ corresponding to an unknown class y, the Bayes rule $d^*: \mathbb{R}^d \mapsto C$ chooses the class $d^*(\boldsymbol{x})$, where

$$d^*(\boldsymbol{x}) \triangleq \arg\max_{c \in C} P_c(\boldsymbol{x}) \qquad P_c(\boldsymbol{x}) \triangleq \Pr\{Y = c|\boldsymbol{X} = \boldsymbol{x}\} \tag{2.1}$$

that is, Bayes rule chooses the class with *maximum a posteriori probability (MAP)*. On occasion we shall also equivalently work with the associated Bayes *discriminant functions* $\delta^*_{i,j} \triangleq P_i - P_j$, $i, j \in C$, so that given a realized feature $\boldsymbol{x}$, by Bayes rule, we would select class $i \in C$ iff $\delta^*_{i,j}(\boldsymbol{x}) > 0$ for all $j \neq i$, $j \in C$. The decision rule and the equivalent discriminant functions induce a corresponding set of M *decision regions* $\{D_c\}_{c \in C}$ where $D_c \triangleq \{\boldsymbol{x} \in \mathbb{R}^d : d^*(\boldsymbol{x}) = c\}$, that is, a partition of $\mathbb{R}^d$ according to the class chosen when the classifier is presented with an input $\boldsymbol{x} \in \mathbb{R}^d$. In relation to these concepts, we are interested in conditions under which:

1. Posterior probability estimates $\tilde{P}_{n,c}$, $c \in C$, constructed from an n-i.i.d. random sample $T_n = \{(\boldsymbol{X}_i, Y_i) \in \mathbb{R}^d \times C\}_{i=1}^n \sim P_{T_n} = \prod_{i=1}^n P_{\boldsymbol{X},Y}$ are (MS) consistent.

[1]See note 1 in Appendix A.3.

2. The consistency of the $\tilde{P}_{n,c}$ implies a corresponding consistency for the *plug-in* or *approximate* Bayes decision rule $\tilde{d}_n$ constructed by substituting the estimated for the actual posterior probabilities, that is,

$$\tilde{d}_n(\boldsymbol{x}) \triangleq \arg\max_{c \in C} \tilde{P}_c(\boldsymbol{x}) \tag{2.2}$$

The approximate Bayes discriminant functions $\tilde{\delta}_n^{i,j}: \mathbb{R}^d \mapsto \mathbb{R}, i, j \in C$, are defined analogously with the estimated posterior probabilities in place of the actual ones. The consistency of a decision rule $d: \mathbb{R}^d \to C$ is defined with respect to its *local risk* $r(d, \boldsymbol{x}) \triangleq \Pr\{d(\boldsymbol{X}) \neq Y | \boldsymbol{X} = \boldsymbol{x}\}$ and its *global risk* $R(d) \triangleq \mathbb{E}[r(d, \boldsymbol{X})]$. The local risk may be viewed as the probability of error as a function of the realized feature vector while the global risk, which averages the probability of classification error over all possible realized feature vectors, can be considered the average probability of classification error. Decision rules with risk converging (in an appropriate mode) to the minimum risk $R^* \triangleq R(d^*)$ achieved by the Bayes rule are known as *Bayes risk consistent (BRC)* (in that mode). Following the point of view in Chapter 1, we shall say BRC when we mean the $L_1(P_{T_n})$ convergence of the *training set conditional* Bayes risk $\mathbb{E}_{X|T_n}[r(d, \boldsymbol{X})|T_n]$ to R^*; other definitions commonly found in the literature are *weak* BRC (when the convergence is i.p.–P_{T_n}) and *strong* BRC (when the convergence is a.s.–P_{T_n}).

2.3 REVIEW OF CURRENT APPROACHES

While the general comments in the Introduction concerning current approaches to RBFN design remain valid, we shall expand the discussion here to include both generic artificial neural network (ANN) and kernel-based design strategies for probability estimation and classification, largely in recognition of the importance attached to those fields in their own right.

2.3.1 Traditional ANN Approaches

As ANN training procedures were (and, for the most part, remain) based on minimizing data-based proxies for the global MSE J_2 [as defined in Eq. (6) in the Introduction], it is not surprising that one approach to an M-class classifier design codes the training targets $Y_i = c \in C$ as the Euclidean unit basis vectors $\boldsymbol{e}_c$ in $\mathbb{R}^M$ so that Y_i is equivalently represented by a vector $\boldsymbol{Y}_i \in \{0, 1\}^M$. In addition to being particularly well-suited for the (0, 1) output range of the sigmoidal activation function commonly used in multilayer perceptron (MLP) networks, this approach, which we shall call the method of *least-squares fitting of indicator functions (LSFI)*, is motivated by the observations that:

1. $\boldsymbol{e}_c$ is the vector of the M class indicator function outputs for the event $Y = c$, that is, $\boldsymbol{e}_c = [I_{\{Y=i\}}(Y = c)]_{i=1}^M$.
2. Regression of a class indicator function yields an L_2 estimate of the corresponding posterior class probability, since

$$P_c(\boldsymbol{x}) = \mathbb{E}[I_{\{Y=c\}}(\boldsymbol{X})|\boldsymbol{X} = \boldsymbol{x}] \qquad (2.3)$$

Note that for each class $c \in C$, we may therefore equivalently write the regression model for P_c as

$$Z_c = P_c(\boldsymbol{X}) + \epsilon_c \qquad c \in C \qquad (2.4)$$

where $Z_c = I_{\{Y=c\}}(\boldsymbol{X}) \in \{0, 1\}$ is the indicator function for class c given input $\boldsymbol{X}$ and $|\epsilon_c| = |Z_c - P_c(\boldsymbol{x})| \leq 1$ for all $\boldsymbol{x} \in \mathbb{R}^d$. By the standard properties of conditional expectation, ϵ_c is uncorrelated with X, $\mathbb{E}[\epsilon_c] = 0$, and $\mathrm{Var}[\epsilon_c] \stackrel{\Delta}{=} \sigma_c^2 \leq 1$.

The initial practical motivations behind the LSFI method have been accompanied by a widely held belief in the ANN community that the method results in MS consistent estimates of posterior probabilities. The line of reasoning involved is as follows:

1. Versions of the (strong) law of large numbers (SLLN) imply the (a.s.) convergence of the proxy $\tilde{R}_2$ to the actual desired MSE cost function J_2 as the size of the training set $n \to \infty$.
2. ANN training procedures that, given a training realization t_n, select the $\tilde{R}_2$-minimizing network parameters $\tilde{\boldsymbol{\theta}}_n$ [and hence network $\tilde{f}(\bullet, \tilde{\boldsymbol{\theta}}_n)$] must equivalently select the J_2-minimizing network parameters as $n \to \infty$.
3. The absolute J_2-minimizing function when Y is the indicator function for a given class is the posterior probability function for that class. Hence ANN training procedures asymptotically (in the number of training data n) yield the J_2-minimizing approximating function $\tilde{f}$ within the approximating function class $\mathcal{F} = \mathcal{F}_{\boldsymbol{\theta}}$.

4. If in addition $\mathcal{F}$ has "sufficient functional capacity" to contain the unknown posterior probabilities, then the ANN output functions recovered asymptotically should be identically equal to the unknown posterior probabilities so that Bayes classification performance is possible.

Several early proofs, for example, those by Hampshire and Perlmutter (1990) and Kanaya and Miyake (1991), essentially follow this model. The work of Ruck et al. (1990) is slightly more guarded in their conclusions as it ends at point 3 and acknowledges that the quality of the approximate Bayes rule depends on the "functional capacity" of the ANN architecture (although the relationship between R_2 and the expected classification error rate is not monotonic, as we shall see). This argument, while suggestive, suffers from a number of deficiencies, many of which were noted in the Introduction in more general terms:

1. Ignoring for the moment the lack of verification of the conditions under which SLLNs hold, step 1 of the argument is incomplete without the key requirement that the a.s. convergence of $\tilde{R}_2(\tilde{f}, T_n)$ to $J_2(\tilde{f})$ is uniform over $\mathcal{F} \ni \tilde{f}$ [see, for example, Section 4.1.2 of White (1989)]. Otherwise, the null set over which the convergence fails can depend on the unknown function f being estimated, an undesirable feature in applications. A proper statement of the desired consistency result with a detailed discussion of the corresponding conditions can be found as Theorem 1 of White (1989).
2. In most ANN applications, one does not usually know *a priori* a *minimally* parameterized function class that is guaranteed to contain the unknown posterior probabilities (otherwise, standard finite parameter estimation methods such as maximum likelihood could be used). On the other hand, for general unknown functions f, for example, continuous f over a known compact set, an infinitely parameterized function class $\mathcal{F}$ is necessary to contain f. Therefore for the antecedent condition of point 4 to hold from the outset, one would have to select a network with an unjustifiably large number of free parameters compared to the amount of training data at hand and risk overfitting, leading to poor generalization as discussed in the Introduction. As Barnard (1992) noted in his comment on Kanaya and Miyake (1991), the notion of "sufficient functional capacity" is largely an impractical one.

A somewhat different approach is taken by Richard and Lippmann (1991) when they assume from the onset that the ANN training procedure minimizes the ensemble average

$$\Delta \triangleq \mathbb{E}\left[\sum_{c=1}^{M}(Y_c - f_c(\boldsymbol{X}, \boldsymbol{\theta}))^2\right] \tag{2.5}$$

where $Y_c \triangleq I_{\{Y=c\}}(\boldsymbol{X}), f_c$ is the ANN output function for class $c \in C$, and $\boldsymbol{X}$ is the feature input RV. With some basic probabilistic manipulations, they arrive at

$$\Delta = \mathbb{E}\left[\sum_{c=1}^{M}(P_c(\boldsymbol{X}) - f_c(\boldsymbol{X}, \boldsymbol{\theta}))^2\right] + \mathbb{E}\left[\sum_{c=1}^{M}\text{Var}(Y_c|\boldsymbol{X})\right] \tag{2.6}$$

Note that this expression could be derived simply by applying the derivation of Eq. (7) in the Introduction to each term of the summation. In any case, since the latter expectation term does not depend on the ANN output functions f_c, they conclude (correctly) that minimizing Δ with respect to $\boldsymbol{\theta}$ is equivalent to minimizing the first term with respect to $\boldsymbol{\theta}$. Clearly, however, this does not imply that *each term* in the former expectation is simultaneously minimized, unless one assumes that the f_c each have effectively independent parameters, which is not usually the case for the reasons given previously with regards to overfitting. Even then, without a proper invocation of an appropriate SLLN, there remains the questionable assumption that ANN training actually selects the network parameters to minimize Δ instead of $\tilde{R}_2$.

In contrast to these incomplete attempts at proving that the ANN-LSFI method yield "estimates" of the Bayes posterior probabilities, Devroye et al. (1996) are able to derive rigorously more general results on the Bayes risk consistency of the LSFI method within the empirical risk minimization framework discussed in the Introduction.[2] Applying the tools of Chapter 1, we offer an alternate route to RBFN-specific consistency results for (posterior) probability estimation and classification. We demonstrate rigorously the consistency of the LSFI method using regularized strict interpolation RBFNs with an a.o. parameter selection technique. We then prove the BRC of the method and show that, from an asymptotic discrimination point of view, the positivity and normalized output constraints often imposed on posterior probability estimates are not

[2]See note 2 in Appendix A.3.

necessary (although, as we shall discuss, they may be desirable under certain circumstances). Furthermore, these theoretical properties of the regularized strict interpolation RBFN design method are proven under precise and (approximately) realizable conditions unlike those often required in the proofs for the other methods.

2.3.2 Kernel-Based Approaches

Given that probability density estimation was a primary motivation behind kernel-based design methods, their advanced development, particularly in theoretical aspects, should come as no surprise. Assuming that a training set T_n is available, posterior probability estimates can be derived in two ways:

1. The *Plug-in KDE Method*: Partition the samples in T_n by class, that is, $T_n = \bigcup_{c\in C} T_{c,n_c}$, $\sum_{c\in C} n_c = n$, where T_{c,n_c} contains only the n_c sample pairs in T_n with $y_i = c$. For each class $c \in C$, form the KDE $\tilde{p}_c$ of the actual *class-conditional* density p_c from the subtraining set T_{c,n_c} in the usual way, that is, each KDE is of the form Eq. (23), albeit not necessarily with the same kernel K or bandwidth sequence $\{h_n\}$. Estimate the corresponding *prior class probability* or *mixing proportion* π_c via, for example, the maximum-likelihood estimator $\tilde{\pi}_c \triangleq n_c/n$ (this estimate is also minimum-variance unbiased). Then applying Bayes rule with the $\tilde{p}_c$ and $\tilde{\pi}_c$ in place of the p_c and π_c yields the estimate $\tilde{P}_c$ of the actual posterior probability P_c of class c as

$$\tilde{P}_c(x) \triangleq \tilde{\pi}_c \tilde{p}_c(x) / \sum_{i\in C} \tilde{\pi}_i \tilde{p}_i(x) \tag{2.7}$$

which may then be used in an approximate Bayes rule.
2. The *KRE Method*: Generate a KRE [typically a Nadaraya–Watson regression estimate (NWRE)] of the posterior class probability for each class $c \in C$ from the M subtraining sets

$$T_{n,c} \triangleq \{(I_{\{Y=c\}}(X_i), X_i) \in \{0, 1\} \times \mathbb{R}^d\}_{i=1}^n \tag{2.8}$$

for use in an approximate Bayes rule. As this method is regression based, it may be considered an LSFI approach.

Among others, Van Ryzin (1966) demonstrates the weak BRC of the plug-in KDE method and derives the order of convergence to consistency under certain conditions on the kernel K and bandwidth sequence $\{h_n\}$. Similarly, Glick (1972) shows that the a.s. pointwise convergence of the density estimates $\tilde{p}_c$, $c \in C$, is sufficient to ensure the a.s. convergence of the sample-based classification error rate to the true classification error rate uniformly over the domain of all decision rules.

For the KRE method, many of the works discussed in the Introduction also include BRC results: Stone (1977) proves the BRC i.p. of his weighted output mean estimators for multiple classification, which is extended by Devroye (1981) in the case of the NWRE to universal BRC a.s. when the bandwidth sequence $\{h_n\}$ satisfies $nh_n^d/\log n \overset{n\to\infty}{\to} \infty$ in addition to the basic NWRE MS consistency conditions (26) and (32) in the Introduction. It is clear that proofs of the BRC of the KRE method carry over to the plug-in KDE method when the latter is consistent as a posterior probability estimate in the same mode that the KRE is, and in this sense there is no essential difference between the two methods. As pointed out, however, in other research, for example, Section 16.2 of Devroye et al. (1996), the corresponding rates to BRC can be arbitrarily loose, in the sense that the convergence (in some appropriate mode) of the parameter and density estimates comprising the plug-in KDE estimate of the posterior probability may be either faster or slower than the convergence of the Bayes risk. This shortcoming is shared with the KRE method, where the MS convergence of the posterior probability estimates implies the BRC of the associated approximate Bayes rule (as will be proven). In a practical sense, however, the KRE LSFI method has an advantage over the plug-in KDE method in that the former uses *all* the available training data to estimate each posterior class probability, while the latter estimates the posterior probability for a given class with only the training data labeled with that class. In other words, even "negative" examples in the form "$\boldsymbol{x}_i$ is not in class c", contribute to the estimate of each posterior class probability using the KRE LSFI method. This choice provides more sample data for each problem and can improve the quality of the resultant estimates, particularly for the *a priori* less probable classes.

Perhaps the closest in spirit to the regularized RBFN framework for estimating posterior probabilities is the method of *inequality-constrained smoothing splines* proposed by Villalobos and Wahba (1987). In their case, by selecting the solution space to be $H(m, d)$, the vector space of d-dimensional functions whose partial derivatives up to total order m (in the distributional sense) are square integrable, and choosing an appropriate penalty functional $H_2 f$ in Eq. (12) in the Introduction, the solution function becomes a linear expansion in multivariate *thin–plate* spline basis functions, that is, functions of the form $K_m(r) = \theta_m r^{2m-d} \ln(r)$ along with the standard multivariate monomials, instead of the typical Gaussian basis function. They then solve a regularized matrix interpolation equation analogous to Eq. (15) in the Introduction except that the solution is subject to a pair of linear equality and inequality constraints, where the latter is an attempt to enforce a [0, 1] range for the posterior probability estimate over a prespecified grid of points. For this purpose, a computational procedure is proposed for a suitably constrained version of GCV (called *GCVC*) for the regularization parameter. Although experimental results with simulated data for a two-dimensional, two-class problem with $m = d = 2$ and a mixture of Gaussian densities indicate visually reasonable modeling of the posterior probabilities, the study does not offer any theoretical results on the consistency for the proposed method. While the (approximate) positive unit range constraint on the solution function is arguably natural when the objective lies solely in the probability estimates themselves, we shall show later that such a range constraint is unnecessary when the objective is discrimination, using the approximate Bayes rules formed from those estimates. Indeed, it is clear that what is *necessary* for classification is the approximation of the *relative* magnitudes of the posterior probabilities, not their *absolute* magnitudes individually. Thus the extra computational complexity introduced into the solution for the optimal function parameters by the positive unit range constraint on the function output range and the subsequent GCVC procedure do not, at least in principle, offer any distinct advantage in classification performance for sufficiently large samples. If direct modeling of posterior probabilities is the goal, penalized likelihood estimates of the *logit* function $\log[P_c/(1 - P_c)]$ which treat the output class variable as a multinomially distributed RV may be considered more appropriate (O'Sullivan et al., 1987).

2.4 THEORETICAL RESULTS FOR REGULARIZED PROBABILITY ESTIMATES

2.4.1 Consistency of Probability Estimates in Mean-Square and Bayes Risk

By Theorem 4 of Chapter 1, regularized strict interpolation RBFN posterior probability estimates designed according to the LSFI method with a.o. regularization parameter sequence are MS consistent whenever the corresponding NWRE is. Because the training data are i.i.d., the conditional error RV $\epsilon(i)$ is no longer homoskedastic, hence of the available scenarios listed in Section 1.4.1, only scenario 3 is potentially relevant. Specifically for that scenario, Andrews (1991) shows that OCV but not GCV satisfies a set of sufficient conditions for $\tilde{L}_2$ and $\tilde{R}_2$–a.o. parameter selection (when conditioned on the training input sequence) in the class of linear estimates. Strictly speaking, since this result is only proven for a discrete parameter space (whose cardinality[3] is permitted to increase with the number of training data), it cannot be automatically assumed to hold for the continuous regularization parameter case. On the other hand, for practical reasons the OCV/GCV cost functions are commonly approximately minimized by discrete sampling of the regularization parameter in any case (for example, see Section 3.7). With these provisos in mind, we may state the following specific instance of Theorem 4:

Theorem 5. *Assume that Theorem 2 holds and let the stationary marginal density p for $\{\mathbf{Z}(i)\}$ be regular in probability. Then the regularized strict interpolation RBFN posterior probability estimates designed according to the LSFI method with regularization parameter sequence $\{\tilde{\lambda}_n\}$ R_2–a.o. as per (1.60) is globally MS consistent whenever the corresponding NWRE is globally MS consistent.*

Proof. By the LSFI method, the output class indicator RV Y is clearly bounded as $|Y| \leq 1$ and the i.i.d. input RV $\mathbf{Z}$ is clearly stationary, so result (1) of Theorem 1 and its extension Theorem 3 hold under our assumed conditions. The remainder of the proof parallels that of Theorem 4. □

From the MS consistency of the regularized strict interpolation RBFN posterior probability estimates, the corresponding BRC is simple to infer from the basic chain of inequalities beginning with the following lemma. The lemma gives a bound relating the expected deviation of posterior probability estimates with the corresponding expected deviation in local risk of the approximate Bayes decision rule formed using those estimates.

Lemma 4 [Györfi (1978)]. Let $\{\tilde{P}_{n,c}\}_{c\in C}$ *be a collection of estimates for a collection of posterior probabilities* $\{P_c\}_{c\in C}$ *and* $\tilde{d}_n$ *be the approximate Bayes rule formed from* $\{\tilde{P}_{n,c}\}_{c\in C}$. *Then*

$$\mathbb{E}[|r(\tilde{d}_n, \mathbf{X}) - r(d^*, \mathbf{X})|] \leq \sum_{c\in C} \mathbb{E}[|\tilde{P}_{n,c}(\mathbf{X}) - P_c(\mathbf{X})|] \tag{2.9}$$

[3]See note 3 in Appendix A.3.

Theorem 6. *In addition to the notation of Lemma 4, let d^* represent the (exact) Bayes rule based on the collection of (true) posterior probabilities $\{P_c\}_{c \in C}$. Then*

$$|R(\tilde{d}_n) - R^*| \leq \sum_{c \in C} \sqrt{\mathbb{E}[|\tilde{P}_{n,c}(X) - P_c(X)|^2]} \tag{2.10}$$

where $R^* \triangleq R(d^*)$ is the Bayes (optimal) risk for the classification problem defined by $\{P_c\}_{c \in C}$.

Proof. Applying Jensen's inequality[4] to the left-hand side of (2.9) gives

$$\mathbb{E}[|r(\tilde{d}_n, X) - r(d^*, X)|] \geq |\mathbb{E}[r(\tilde{d}_n, X) - r(d^*, X)]| = |R(\tilde{d}_n) - R^*| \tag{2.11}$$

while the Cauchy-Schwarz inequality implies that the right-hand side satisfies

$$\mathbb{E}[|\tilde{P}_{n,c}(X) - P_c(X)|] \leq \sqrt{\mathbb{E}[|\tilde{P}_{n,c}(X) - P_c(X)|^2]} \tag{2.12}$$

Combining the two inequalities gives the desired result. □

Theorem 6 bounds the rate to BRC as being no slower than the square-root of the minimum rate to the R_2-consistency amongst the corresponding posterior class probability estimates. Results exist, for example, Theorem 6.5 of (Devroye et al., 1996), which state that the *ratio* of left-hand side of (2.10) to its bound on the right-hand side vanishes (as $n \to \infty$) in the two-class case for the risk of any approximate Bayes rule formed from (weakly) MS convergent posterior probability estimates. In fact, if the Bayes risk R^* is zero, that is, in cases where perfect classification is (in principle) possible, $|R(\tilde{d}_n) - R^*|$ vanishes as fast as the *square* of the corresponding rates to MS consistency of the posterior probability estimates. Devroye et al. (1996) also give an example of a case where convergence of an approximate Bayes rules' risk to the Bayes risk occurs *without* the MS consistency of the class posterior probability estimates. Taken together, these results reinforce the point that while MS consistency of posterior probability estimates is certainly *sufficient* to guarantee a worst-case rate to BRC for the corresponding approximate Bayes rule,

[4]See note 4 in Appendix A.3.

such consistency is not by any means necessary and, as is typical of such worst-case rates, may be of more theoretical than practical interest.

We should also point out another way of proving the BRC of regularized strict interpolation RBFN posterior probability estimates with a.o. regularization parameter sequence. This approach shows that the convergence i.p.–$P_{\mathbf{X},T_n}$ of the approximate Bayes discriminant functions to the actual Bayes discriminant functions is sufficient to ensure the convergence i.p.–$P_{\mathbf{X},T_n}$ of their classification decisions and hence risk or error rates. As discussed previously, this result implies that the positive unit range constraint on the posterior probability estimate outputs imposed by the method of Villalobos and Wahba (1987) is not necessary for asymptotically consistent classification, only the replication of the signs of the discriminant functions are. In this restricted sense, our result supports the intuition "classification is easier than regression" that has sometimes been enunciated in pattern recognition circles, for example, in Section 6.7 of Devroye et al. (1996). We begin by proving the two-class case as a lemma and then extend it to the multiclass case in a subsequent theorem.

Lemma 5. *Let $\tilde{P}_{n,1}$ and $\tilde{P}_{n,2}$ be globally MS consistent estimates of posterior probabilities P_1 and P_2, respectively, according to the conditions of Theorem 4. Assume further that $\tilde{P}_{n,1} \neq \tilde{P}_{n,2}$ a.s.–$P_{\mathbf{X},T_n}$ and $P_1 \neq P_2$ a.s.–$P_{\mathbf{X}}$. Let $\tilde{d}$ (d^*) be the approximate (exact) Bayes decision rule formed from $\tilde{P}_{n,1}$ and $\tilde{P}_{n,2}$ (P_1 and P_2). Then*

$$\lim_{n\to\infty} P_{\mathbf{X},T_n}(\{\mathbf{x}\in\mathbb{R}^d,\, t_n\in\tau^n: \tilde{d}(\mathbf{x}) \neq d^*(\mathbf{x})\}) = 0 \tag{2.13}$$

Proof. For $c\in\{1,2\}$, since we have assumed that $R_2(P_c,\tilde{P}_{n,c}) \overset{n\to\infty}{\to} 0$, we also have the convergence in probability

$$|\tilde{P}_{n,c}(\mathbf{X}) - P_c(\mathbf{X})| \overset{n\to\infty}{\to} 0 \text{ i.p.--} P_{\mathbf{X},T_n} \tag{2.14}$$

which implies the same convergence of corresponding approximate Bayes discriminant function $\tilde{\delta}_n \triangleq \tilde{P}_{n,1} - \tilde{P}_{n,2}$ to the actual Bayes discriminant function $\delta^* \triangleq P_1 - P_2$. Hence $D^* \triangleq \{\mathbf{x}\in\mathbb{R}^d: \delta^*(\mathbf{x}) > 0\}$ and $\tilde{D}_n \triangleq \{\mathbf{x}\in\mathbb{R}^d,\, t_n\in\tau^n: \tilde{\delta}_n(\mathbf{x}) > 0\}$ are the actual and approximate Bayes decision regions, respectively, for selecting class 1 over class 2. The desired result (2.13) is equivalently stated

$$\lim_{n\to\infty} (P_{\mathbf{X},T_n}(D^* \,\Delta\, \tilde{D}_n) + P_{\mathbf{X},T_n}(D^{*c} \,\Delta\, \tilde{D}_n^c)) = 0 \tag{2.15}$$

Let us consider the first probability term, as symmetry will allow us to apply the subsequent argument to the second probability term with only minor changes. By additivity for disjoint events,

$$P_{X,T_n}(D^*\Delta\tilde{D}_n) = P_{X,T_n}(D^* \cap \tilde{D}_n^c) + P_{X,T_n}(D^{*c} \cap \tilde{D}_n) \tag{2.16}$$

For $(\boldsymbol{x}, t_n) \in (D^* \cap \tilde{D}_n^c) \cup (D^{*c} \cap \tilde{D}_n)$, we have, by the definitions of δ^* and $\tilde{\delta}_n$, the implication

$$|\delta^*(\boldsymbol{x}) - \tilde{\delta}_n(\boldsymbol{x})| < \epsilon \Rightarrow |\delta^*(\boldsymbol{x})| + |\tilde{\delta}_n(\boldsymbol{x})| < \epsilon \tag{2.17}$$

Letting $s_n \triangleq |\delta^*| + |\tilde{\delta}_n|$ and $S_{n,\epsilon} \triangleq \{\boldsymbol{x} \in \mathbb{R}^d,\ t_n \in \tau^n : s_n(\boldsymbol{x}) < \epsilon\}$, this implication is equivalently stated

$$(D^* \cap \tilde{D}_n^c) \cup (D^{*c} \cap \tilde{D}_n) \subseteq S_{n,\epsilon} \tag{2.18}$$

so that

$$P_{X,T_n}(D^* \cap \tilde{D}_n^c) + P_{X,T_n}(D^{*c} \cap \tilde{D}_n) \leq P_{X,T_n}(S_{n,\epsilon}) \tag{2.19}$$

For any given n, we claim that $P_{X,T_n}(S_{n,\epsilon})$ can be made arbitrarily small by choosing ϵ sufficiently small. To see this, take $\epsilon = 1/m$ and let $A_m \triangleq S_{n,1/m}$ (for fixed n). Because $A_1 \supset A_2 \supset \cdots \supset A_m \supset \cdots$ forms a monotone decreasing sequence of events, by continuity in probability, we have that

$$\lim_{m\to\infty} P_{X,T_n}(A_m) = P_{X,T_n}\left(\bigcap_{i=1}^{\infty} A_i\right) \tag{2.20}$$

The right-hand side $\bigcap_{i=1}^{\infty} A_i = A_0 \triangleq \{\boldsymbol{x} \in \mathbb{R}^d,\ t_n \in \tau^n : s_n(\boldsymbol{x}) = 0\}$ is precisely the set of points where neither the discriminant function δ nor $\tilde{\delta}_n$ yield an unambiguous decision, a set which by assumption has null P_{X,T_n}. Thus the claim is verified so that the first probability term on the right-hand side of (2.16) can be made arbitrarily small if n is sufficiently large. This same argument applies to the second probability term on the right-hand side of (2.16) by exchanging D and $\tilde{D}_n$ with D^c and $\tilde{D}_n^c$, respectively, and replacing δ and $\tilde{\delta}_n$ with $-\delta$ and $-\tilde{\delta}_n$, respectively. With these results in place, given any $\gamma > 0$, we can choose n sufficiently large so that

$$\max(P_{X,T_n}(D^* \Delta \tilde{D}_n), P_{X,T_n}(D^{*c} \Delta \tilde{D}_n^c)) < \gamma/2 \tag{2.21}$$

whence (2.13) is proven. □

Theorem 7. *Let $\{\tilde{P}_{n,c}\}_{c\in C}$ be a collection of globally MS consistent estimates for a corresponding collection of M posterior probabilities $\{P_c\}_{c\in C}$, according to the conditions of Theorem 4. Assume further that for any $i, j \in C$ with $i \neq j$, $\tilde{P}_{n,i} \neq \tilde{P}_{n,j}$ a.s.–P_{X,T_n} and $P_i \neq P_j$ a.s.–P_X. Let $\tilde{d}$ (d^*) be the approximate (exact) Bayes decision rule formed from $\{\tilde{P}_{n,c}\}_{c\in C}$ ($\{P_c\}_{c\in C}$). Then*

$$\lim_{n\to\infty} P_{X,T_n}(\{\mathbf{x}\in\mathbb{R}^d, t_n\in\tau^n: \tilde{d}_n(\mathbf{x}) \neq d^*(\mathbf{x})\}) = 0 \tag{2.22}$$

and, consequently,

$$\lim_{n\to\infty} R(\tilde{d}_n(\mathbf{x})) = R^* \tag{2.23}$$

that is, the approximate Bayes decision rules are BRC.

Proof. For the multiclass case, we have the basic bound

$$P_{X,T_n}(\{\mathbf{x}\in\mathbb{R}^d: \tilde{d}(\mathbf{x}) \neq d^*(\mathbf{x})\}) \\ \leq P_{X,T_n}\left(\bigcup_{i\in C}\ \bigcup_{j\in C, j\neq i} \{\mathbf{x}\in\mathbb{R}^d, t_n\in\tau^n: d^*_{i,j}(\mathbf{x}) \neq \tilde{d}_n^{i,j}(\mathbf{x})\}\right) \tag{2.24}$$

where $d^*_{i,j}$ and $\tilde{d}_n^{i,j}$ are the actual and approximate Bayes rules, respectively, for the two-class subproblem defined by P_i and P_j. The inequality arises from the fact that if the actual and approximate Bayes decision rules for the full M-class problem disagree for a given $\mathbf{x}\in\mathbb{R}^d$, then $d^*_{i,j}(\mathbf{x}) \neq \tilde{d}_n^{i,j}(\mathbf{x})$ for at least one pair of classes i and j with $i \neq j$, from the $N_2 \triangleq \binom{M}{2} = M(M-1)/2$ possible two-class subproblems over C. Given $\epsilon > 0$ and applying the union bound to the right-hand side of (2.24), we see that by choosing n sufficiently large so that

$$\max_{\substack{i,j\in C\\ i\neq j}} P_{X,T_n}(\{\mathbf{x}\in\mathbb{R}^d, t_n\in\tau^n: d^*_{i,j}(\mathbf{x}) \neq \tilde{d}_n^{i,j}(\mathbf{x})\}) \leq \frac{\epsilon}{N_2} \tag{2.25}$$

we satisfy (2.22). To show Theorem 7 using this result, let $\tilde{E}_n \triangleq \{(\mathbf{x}, y)\in\mathbb{R}^d\times C, t_n\in\tau^n: \tilde{d}_n(\mathbf{x}) \neq y\}$ and $E^* \triangleq \{(\mathbf{x}, y)\in\mathbb{R}^d\times C: d^*(\mathbf{x}) \neq y\}$

be the classifier error events for the approximate and actual Bayes decision rules, respectively. Since

$$|P_{X,Y,T_n}(\tilde{E}_n) - P_{X,Y}(E^*)| \leq P_{X,Y,T_n}(\tilde{E}_n \,\Delta\, E^*) \tag{2.26}$$

and because the event $\tilde{E}_n \,\Delta\, E^*$ occurs if and only if exactly one of the two decision rules commits an error, it is a subset of the event that the two decision rules disagree. But we have just proven in (2.22) that this disagreement probability can be made arbitrarily small, thus completing the proof of Theorem 7. □

2.5 SUMMARY AND DISCUSSION

The connection between regression on class indicator functions and classification by Bayes rule with posterior probabilities was reviewed. This connection motivates several common ANN arguments for the Bayes risk consistency of approximate Bayes rules constructed from ANN estimates of posterior probabilities via least-squares fitting of indicator functions. We then showed how the deficiencies in these ANN arguments for BRC can be avoided by applying the previously developed results for regularized strict interpolation RBFNs with asymptotically optimal regularization parameter sequence. Specifically, the global MS consistency of such regularized strict interpolation RBFN posterior probability estimates follows naturally as a special case of Theorem 4. The BRC of the approximate Bayes rule classifiers using these MS consistent regularized strict interpolation RBFNs is then deduced from a theorem bounding the deviation from optimal Bayes risk in terms of the class posterior probability MS estimation errors. In addition to this principal result, the following related developments were also explored:

1. In the alternative proof given in Theorem 7 for the BRC of approximate Bayes decision rules based on MS approximations of posterior probabilities, we have demonstrated that the approximate and actual Bayes decision rules agree with probability approaching one without any requirement that the MS approximations satisfy the positive unit range constraint of the actual posterior probabilities. Instead, MS consistent posterior probability estimates are shown to result in corresponding i.p.-consistent approximations of the Bayes decision regions for each class, which is the minimum requirement

for BRC classification. On the other hand, a plausible argument could be made that imposing such positivity and normalized output constraints on the posterior probability estimates should have some practical benefits, for example, a more rapid MS convergence to the actual posterior probabilities. While the heuristics behind this notion are clear, we are not aware of any theoretical evidence supporting it. Furthermore, given the results of (Devroye et al., 1996) regarding the rather weak connection between the rate to MS consistency for posterior probability estimates and rate to BRC for the corresponding approximate Bayes decision rule, it is uncertain what value such theoretical evidence would have, if obtained.

2. To establish global BRC, we require that the density for the feature RV $\boldsymbol{X}$ satisfy the regularity conditions discussed in Section 1.4.2 so that Theorem 4 may hold. Often in the pattern recognition literature one deals with the class-conditional densities $\{p_c\}_{c \in C}$, where p_c is the density of $\boldsymbol{X}|Y = c$, for example, see the description of the plug-in KDE method. Since the (unconditional) feature density p is related to the $\{p_c\}_{c \in C}$ by $p = \sum_{c \in C} \pi_c p_c$, where the $\{\pi_c\}_{c \in C}$ are the prior class probabilities, p is regular in probability if the $\{p_c\}_{c \in C}$ are.
3. While in principle GCV(C) is not provably a.o. for continuous parameter selection with heteroskedastic errors as in the LSFI approach, Villalobos and Wahba (1987) report little loss in modeling performance for their simulated data using the "plain", that is, non-heteroskedasticity corrected, GCV procedure versus a weighted GCV procedure which does correct for the heteroskedasticity of the error (as elaborated in Section 1.4.1). This result accords with our general experience in the experiments for this monograph and has been corroborated by others, for example, Section 4.9 of (Wahba, 1990). A more theoretically sound approach is to use the "ordinary" or leave-out-one CV procedure for parameter selection, although even this ameliorative measure has its limitations as indicated earlier.

We have now formally established the BRC of the approximate Bayes decision rules for MS consistent posterior probability estimates via the regularized strict interpolation RBFN method with a.o. parameter selection procedure. The verification and accompanying discussion substantiate the long-standing belief within the ANN community that the LSFI approach can lead to BRC-approximate Bayes decision rules while pointing out some of the pitfalls and limitations of the LSFI methods which are not as well-known as they ought to be. Even though such theoretical results do not imply that the LSFI method is the most appropriate choice in general situations, their existence does at least support their application to problems for which the method is sufficient to achieve the desired results.

3

NONLINEAR TIME-SERIES PREDICTION

3.1. INTRODUCTION

By a time series, we usually mean a pair of discrete-time stochastic process $\{(\boldsymbol{Z}(i), Y(i)) \in \mathbb{R}^d \times \mathbb{R}\}_{i=0}^{\infty}$. Time series that are in principle deterministic, for example, *chaotic* time series, can be handled as stochastic processes with singular distributions or, more realistically, as processes with a deterministic component contaminated by additive noise. In applications, the vector input process $\{\boldsymbol{Z}(i)\}$ is often composed from one or more (observable) scalar processes $\{X_j(i)\}$, $j = 1, 2, \ldots, d$, via $\boldsymbol{Z}(i) \stackrel{\Delta}{=} [X_j(i)]_{j=1}^{d}$. A particularly important construction of this kind is the *autoregressive (AR)* case in which $Y(i) \stackrel{\Delta}{=} X(i)$, $X_j(i) \stackrel{\Delta}{=} X(i-j)$ are the *delay inputs*, and $\boldsymbol{Z}(i)$ is the *delayed input vector*. Here d is called the *autoregressive* or *delayed input order* and represents the length of dependence between the present and the immediate past of the process $\{X(i)\}$. An AR time series with unbounded input order is in principle possible but, as a practical measure, we shall normally assume that d is effectively finite.

In this chapter, we examine the issue of nonlinear time-series prediction using radial basis function (RBF) artificial neural networks (ANNs) and related kernel regression estimates (KREs). After briefly surveying the existing approaches and results in this area, we focus on the class of nonlinear AR (NLAR) time series generated by an i.i.d. noise process. Specifically, we show how the previously developed results on the consistency of regularized strict interpolation RBFNs (with asymptotically optimal regularization parameter sequence) for stationary input densities carry over to the nonstationary but geometrically ergodic processes that characterize the NLAR prediction problem we consider. We also develop infinite and finite-memory recursive update procedures for the RBFN weights analogous to the standard linear recursive least-squares (RLS) update. Finally, we present results on the nonlinear one-step-ahead prediction of speech signals using dynamically per-step updated regularized strict interpolation RBFNs with asymptotically optimal regularization parameter sequences.

3.2 PROBLEM DESCRIPTION

The *filtering* or *estimation* problem is to estimate $Y(i)$ given $\boldsymbol{Z}(i)$ with minimum J_2 risk, that is, minimum mean-square error. The *prediction* problem can be viewed as a case of the filtering problem in which $\boldsymbol{Z}(i)$ is a delayed input vector, that is, contains information only up to (and including) time step $i-1$. It is well known that the optimum estimator in this situation is the conditional mean $f_i(\bullet) \stackrel{\Delta}{=} \mathbb{E}[Y(i)|\boldsymbol{Z}(i) = \bullet]$. Such optimal MMSE estimators are well-developed for the case where $\{Y(i)\}$ and $\{\boldsymbol{Z}(i)\}$ are jointly weakly (or wide sense) stationary (WSS) Gaussian processes, in which case f_i is a linear function, and for cases where $\{Y(i)\}$ and $\{\boldsymbol{Z}(i)\}$ are known to be related by a given parametric form. When neither of these conditions necessarily obtains, the nonparametric ANN and KRE approaches once again generate an estimate $\tilde{f}$ of f under the stochastic or noisy data models, as appropriate. The estimate $\tilde{Y}(i)$ is then formed by the plug-in estimate $\tilde{Y}(i) \stackrel{\Delta}{=} \tilde{f}(\boldsymbol{Z}(i))$. The specific nomenclature of time series aside, the distinguishing characteristics of the time-series estimation problem are:

1. The possibility that the various quantities of interest, for example, the statistics of the input and output processes, may fluctuate with time, that is, *nonstationarity.*
2. The possibility that the training data may be dependent from sample to sample, that is, *dependence.*

as opposed to the i.i.d. assumptions under which functional estimation problems such as probability estimation and classification are overwhelmingly posed. Note that these two conditions need not occur simultaneously, which is somewhat fortuitous since dealing with both of these added complications together would indeed be a challenging task. For the most part, we shall content ourselves with specific cases of each condition individually to allow tractable analysis.

3.3 REVIEW OF CURRENT APPROACHES

The comments in the Introduction concerning current approaches to RBFN design generally apply in the area of time-series estimation, as they did in the probability estimation problem, except that in this case we shall focus our attention on only those approaches that explicitly relate to RBFNs.

3.3.1 Traditional ANN Approaches

A survey of the literature indicates that time-series prediction may be considered one of the original applications for RBFN. Early design methods are characterized by their assumption of stationarity and reliance on heuristics, often to improve prediction performance or reduce computational complexity. For example, when Casdagli (1989) applies strict interpolation RBFNs (without regularization) to the autoregressive prediction of chaotic time series, the computational complexity of solving the strict interpolation (SI) problem [Eq. (15) in the Introduction] exactly for large n is reduced by considering only the 50 inputs in the training set t_n closest (in Euclidean norm) to a given input training point $z(i)$ when approximating its target value $y(i)$, resulting in a predictor with a piecewise rather than global smoothness normally associated with RBFNs. Broomhead and Lowe (1988) consider similar chaotic modeling problems for the doubling and quadratic maps using their pseudo-inverse method as outlined in the Introduction; network parameters such as the number of centers and the kernel bandwidth are chosen *ad hoc*, except for the network centers, which are chosen to be uniformly spaced within the known unit range of the maps.

Keeping with the autoregressive prediction of chaotic and cyclical time series, He and Lapedes (1993) propose a *successive approximation* approach in which the available training data are partitioned into disjoint subsets, each of which is used to provide the centers of a corresponding RBF subnetwork trained with the pseudo-inverse method to approximate the overall (common) training set. The outputs of these RBF subnetworks are then assigned linear weights via the standard least-squares pseudo-inverse approximation to the overall training set. Compared to the usual approach of designing a single strict interpolation RBFN using all the available data as centers and training targets, it is difficult to see what the provable advantages of this two-stage least-squares approximation design technique are, other than the lower computational complexity and lessened likelihood of singularity in the interpolation matrices for cyclical time series as claimed by the authors.

Although these pseudo-inverse-based approaches are motivated by the desire for lower computational complexity in design and better generalization by avoiding overfitting in the presence of noise, they offer neither any specific theory with which to select the network centers from the data nor do they indicate what sort of noise-immunity can be expected as the number of centers varies. More generally, these methods also do not explicitly address the issues of noisy/stochastic data models or possible correlation and nonstationarity in the time series data.

Another route to the design of RBFNs for statistical time-series prediction lies in the point of view that the output time series is linear in a number of unknown *state variables* that are assumed *a priori* to be related to the delayed input vector (of a known order) in the observable inputs via a known radial kernel. In such a case, if the output process is assumed to contain additive white Gaussian noise so that the optimal linear weights are posteriorly Gaussian distributed, one may, as Terano et al. (1992) do, apply the standard linear Kalman filter (which, in this case, reduces to the RLS algorithm) to recursively estimate the required weights. Note that in this approach, although the centers of the RBFN naturally follow changes in the input time series, there is still no guarantee of suitability for nonstationary systems as the weight update algorithm assumes weak stationarity for all processes involved.

More recent efforts have been directed toward the development of principled time-adaptive RBF networks so that both nonstationary and stationary processes may be *tracked* on an ongoing basis. One example is Kadirkamanathan and Kadirkamanathan (1996), who extend the work of Terano et al. (1992) by using an *extended* version of the RLS algorithm that allows the optimal state-space weights $\mathbf{w}^*(i)$ to drift according to a random walk $\mathbf{w}^*(i+1) = \mathbf{w}^*(i) + \mathbf{e}(i)$, where $\mathbf{e}(i)$ is a Gaussian white noise process. For *modally* nonstationary time series, that is, time series generated by piecewise constant switching among a fixed number of state-space mappings, and first-order Markovian transition between modes, they further use a *multiple model algorithm* to select (via Bayes inference) the "best" predictor from a number of candidate models running in parallel. A similar application of Bayesian inference in the nonstationary case also figures prominently in (Lowe and McLachlan, 1995). There, however, *arbitrary* nonlinear state-space mappings, that is, those not necessarily in the linear span of the chosen basis functions, are accommodated by extended Kalman filters of second and higher order that produce recursive Bayes estimates of the RBF network weights that best approximate (in mean square) the nonlinear mapping.

Compared with the earlier pseudo-inverse-based design techniques, these more principled approaches can account for both nonstationarity and dependence in the time series data; the trade-off is, of course, the need for stronger assumptions regarding the time series model, for example, that the output process is a nonlinear function of the input process with additive Gaussian noise of a known covariance. Physical intuition can justify such assumptions in some situations but it would certainly be desirable if optimal estimates could be obtained without them.

3.4 KERNEL REGRESSION APPROACH

With their localized smoothing akin to filtering, KRE finds natural application to statistical time series. Early motivation for this approach to time series analysis arose from the desire to apply the KDE to (strictly) stationary processes with dependent samples rather than i.i.d. samples, as is usually assumed (Földes, 1974; Bosq, 1973, 1975; Pham and Tran, 1991). From this point, study developed naturally for the Nadaraya–Watson regression estimate (NWRE) as both a *function* estimate as well as a *plug-in* estimate described previously. Surveys may be found in Györfi et al. (1989) and, more recently, in Bosq (1996). For both, the primary focus is on the problem of estimating the stationary regression function $f(\bullet) = \mathbb{E}[Y(i)|\boldsymbol{Z}(i) = \bullet]$, $\forall i$, in the case of a stationary input density under a mixing condition. Györfi et al. (1989) show and provide rates for the a.s. uniform consistency of the NWRE as a function estimate over compact sets in $\mathbb{R}^d$ under ϕ, ρ, and α-mixing conditions. For stationary Markov processes of order $q < \infty$ under conditions leading to geometrically ϕ mixing (GPM), they also demonstrate the a.s. consistency in absolute value of the plug-in NWRE autoregressive predictor by comparison to an existing a.s. uniform consistency result over compact sets.

Finally, an a.s. uniform consistency result for the NWRE as an autoregressive function estimate for stationary ergodic processes with continuous conditional densities is stated, although no rates are given. Considering the NWRE as a function estimate when the input process is geometrically strong (or α) mixing (GSM), Bosq (1996) proves its pointwise mean-square (MS) consistency over $\mathbb{R}^d$ and its a.s. uniform consistency over regular sequences of sets [see Eq. (1.64)]. Estimates of the rate of convergence for the plug-in NWRE predictor quadratic and absolute error in the case of stationary fixed-order Markov processes are proven, along with a brief note on the extension to general stationary processes. In all cases, conditions akin to those in the corresponding kernel density estimation (KDE) cases are (not surprisingly) required to hold.

Actually, as developed in Györfi et al. (1989), the assumption of stationary input density is necessary only if rates of convergence are required; otherwise, a condition such as (A.1) discussed preceding Theorem 1 (see Chapter 1) is sufficient to ensure convergence. Thus, so long as the regression function f is fixed, situations in which the nonstationarity in the output process $\{Y(i)\}$ is caused by a nonstationary input process $\{\boldsymbol{Z}(i)\}$ are permitted. Beyond this general case, the first survey goes on to examine specific cases of nonstationarity that remain within the regression framework developed. A particularly interesting one, which we shall consider further on, is the class of autoregressive processes of order $d \geq 1$ generated by i.i.d. innovations, that is,

$$X(i) = f(\boldsymbol{X}_d(i-1)) + \epsilon(i), \qquad i = 1, 2, \ldots \tag{3.1}$$

where $\boldsymbol{X}_d(i) \triangleq [X(i-j)]_{j=0}^{d-1}$ and $\{\epsilon(i)\}$ is an i.i.d. noise process with zero mean, bounded variance σ^2, and independent of the initial state vector $\boldsymbol{X}_d(0)$. The process $\{X(i)\}$ is clearly dependent and Markov with nonstationary marginal input density. It can be shown that if f is bounded and the probability law of $\{\epsilon(i)\}$ is absolutely continuous with respect to Lebesgue measure, then $\{X(i)\}$ is GPM for which the NWRE is a.s.–P_{T_n} convergent in absolute value.[1] The second survey, however, considers only a simple form of nonstationarity in which

$$Z(i) = X(i) + s(i) \qquad i \in \mathbb{Z} \tag{3.2}$$

where $\{X(i)\}$ is an unobserved, scalar real-valued strictly stationary process and $\{s(i)\}$ is a deterministic sequence. Under some conditions

[1]This result is proven as Theorem 3.4.11 of Györfi et al. (1989).

on the perturbation induced by $\{s(i)\}$ on the marginal density p of $\{X(i)\}$ and the joint density $p_{0,1}$ of $(X(0), X(1))$, it is demonstrated that the NWRE of the autoregression $E[Z(i)|\boldsymbol{Z}_i(i)]$ is a.s. consistent in absolute value with rate n^{δ}, where $\delta \geq 0$ is determined by the perturbation conditions. In effect, this result states that the NWRE exhibits some robustness to mild forms of nonstationarity.

A related area involving nonstationary processes to which the NWRE has been applied is the *identification* of nonlinear systems. Rutkowski (1985a) considers identifying in the limit a measurable function $m: \mathbb{R}^d \to \mathbb{R}$ from a noisy training set T_n satisfying

$$Y(i) = f_i(\boldsymbol{Z}(i)) + \epsilon(i), \quad i \in \mathbb{N} \tag{3.3}$$

where $\{\boldsymbol{Z}(i)\}$ is an i.i.d. process with density p and independent of the i.i.d. zero-mean, finite variance noise process $\{\epsilon(i)\}$. Under these conditions, we see that f_i is the regression function of $Y(i)$ with respect to $\boldsymbol{Z}(i)$, and the nonstationarity of the output process $\{Y(i)\}$ is due solely to the time variation of the regression function. Then, if one imposes the *quasi-stationarity* condition

$$\lim_{i \to \infty} \sup_{z \in \mathbb{R}^d} |f_i(z) - f(z)| = 0 \tag{3.4}$$

such that $f_i \cdot p$ converges (as $i \to \infty$) to $f \cdot p$ either uniformly over $\mathbb{R}^d$ or in $L_p(\mathbb{R}^d, P_{T_n})$ norm sufficiently fast for $p = 1$ or 2, then the NWRE $\tilde{f}_n$ of f_n based on T_n is pointwise consistent i.p. or a.s.–P_{T_n}, depending on the conditions placed upon the bandwidth and kernel. Note, however, that the number of kernel functions in the NWRE must grow in step with the size of the training set to achieve such consistency, which makes this result principally of theoretical interest. This shortcoming is alleviated, however, when Rutkowski (1985b) presents a *recursive* version of the aforementioned algorithm to *track* f_i, which need not be convergent in the sense of (3.4). What is salient about this recursion is that the contribution $a(n+1) > 0$ of a new training input datum $z(n+1)$ to $\tilde{f}_n$ via its kernel function $K((\bullet - z(n+1))/h_n)$ asymptotically vanishes [but not too quickly, as $\sum_{i=1}^{\infty} a(i) = \infty$], that is, the function estimate is, for practical purposes, of fixed size when n is sufficiently large. Among one of several conditions, if the bandwidth and kernel for the NWRE are such that the KDE $\tilde{p}_n$ is pointwise consistent i.p. or a.s.–P_{T_n} with respect to p, the stationary input density, then this algorithm is also pointwise consistent in the same mode.

Once again, compared to the ANN RBF approaches, the KRE-based approaches offer stronger theoretical support for their design decisions. The issue of dependent training data arising in correlated time series is addressed by adapting existing KDE results for the same, while the more general question of nonstationarity is analyzed for certain important cases. With some modifications, it is expected that the regularized strict interpolation RBFN approach can exploit this body of knowledge to demonstrate its consistency and asymptotic R_2 optimality (which in this case translates into MS prediction error), and that is precisely the topic for the forthcoming sections.

3.5 CONSISTENCY OF PREDICTION

For time series with stationary input density and dependence described by a GSM condition, the relevant MS consistency results from Bosq (1996) for the NWRE as a function estimate and a plug-in predictor carry over directly [by application of Theorem 4 (see Chapter 1)] to the regularized strict interpolation RBFN with a.o.-selected regularization parameter. If the input process does not possess a stationary density, there are two possible solutions:

1. We may construct modified versions of the constructive approximation theorems in Section 1.2, which use condition (A.1) in place of the assumption of a stationary marginal input density. With this change, nonstationary input processes would be admissible in the function estimation context, so long as the underlying regression function is fixed.
2. For the specific case of prediction in the Markovian autoregressive model (3.1), it is known that if f is bounded and $\{\epsilon(i)\}$ is an i.i.d. zero-mean, finite variance process with measure absolutely continuous with respect to Lebesgue measure, then both the scalar process $\{X(i)\}$ defined by (3.1) and the *equivalent vector process* $\{\boldsymbol{X}_d(i)\}$ defined by

$$\begin{aligned}\boldsymbol{X}_d(i) &= [f(\boldsymbol{X}_d(i-1)), X(i-1), X(i-2), \ldots, X(i-d+1)]^{\top} \\ &\quad + \epsilon(i)\boldsymbol{e}_1 \\ &\triangleq T(\boldsymbol{X}_d(i-1)) + \epsilon(i)\boldsymbol{e}_1 \qquad (3.5)\end{aligned}$$

where $e_1 \triangleq [1, 0, \ldots, 0]^{\top}$ is the first unit vector in $\mathbb{R}^d$, are GPM, and therefore also GSM (since ϕ mixing implies α mixing).[2] In other words, the dependence induced by the Markov autoregressive construction satisfies the mixing conditions sufficient for the approximation theorems of Section 1.2. To deal with the remaining issue of input process stationarity, we may impose conditions so that the autoregressive process is *geometrically ergodic*, that is,

$$|P_n(P_0) - \pi|_V = \mathcal{O}(\rho^n) \text{ for some } \rho \in (0, 1) \tag{3.6}$$

where P_n is the (marginal) probability measure for $\boldsymbol{X}_d(n)$, P_i (P_j) for $i \geq j$ denotes the probability measure for $\boldsymbol{X}_d(i)$ given that $\boldsymbol{X}_d(j)$ has probability measure P_j, and $\|\bullet\|_V$ denotes the total variation norm for the space $\mathcal{L}$ of probability measures over $\mathcal{B}(\mathbb{R}^d)$, that is, for two probability measures $P, Q \in \mathcal{L}$,

$$\|P - Q\|_V \triangleq \sup_{B \in \mathcal{B}(\mathbb{R}^d)} |P(B) - Q(B)| \tag{3.7}$$

Here π denotes the *stationary* or *invariant* measure of the continuous $\mathbb{R}^d$-valued Markov chain (3.5). The idea of geometric ergodicity is that regardless of the distribution P_0 of the initial state $\boldsymbol{X}_d(0)$, the state of the Markov chain approaches stationarity exponentially fast. Hence, applying (3.6) with n sufficiently large, the previously developed approximation theorems for the stationary marginal input density case hold after replacing the common input measure P and density p with the invariant measure π and density p_π, respectively. An exposition on the conditions for the general continuous vector-valued version of (3.5) to be geometrically ergodic may be found in Chapter 4 and Appendix 1 of Tong (1990) and are summarized in Section 2.4.0.2 of Douhkan (1994). Adapting the results for the general case to the equivalent autoregressive models (3.1) and (3.5), we find that sufficient conditions are:

a. $\{\epsilon(i)\}$ satisfies $\mathbb{E}[|\epsilon(i)|] < \infty$ and has an everywhere continuous and positive density with respect to Lebesgue measure.
b. f is Lipschitz (and hence bounded over bounded sets) in $\mathbb{R}^d$, has $f(\mathbf{0}) = 0$ [so that $T(\mathbf{0}) = \mathbf{0}$], and is *exponentially asymptotically stable in the large*, that is, $\exists A, c > 0$ such that $\forall n \in \mathbb{Z}^+$ and

[2] This result is proven as Theorem 3.4.10 of Györfi et al. (1989).

$x(0) \in \mathbb{R}^d$, $\|x(n)\| \le A \exp(-cn)\|x(0)\|$, where $x(n) \triangleq T^n(x(0))$ is the n-fold composition of T applied to $x(0)$.

Of the two conditions, the second is obviously the more restrictive one because it requires that the underlying mapping f satisfy a strong contractivity condition (although it does allow the stable point of the map T to be other than $\mathbf{0}$ by applying a suitable translation). Exponential decay in transiently driven physical systems is quite plausible, however, which implies that the exponentially asymptotic stability condition may hold at least locally within a given time series.

Since the Markovian autoregressive model Eq. (3.1) plays a central role in our speech prediction application, we shall follow the second approach in dealing with input process nonstationarity. In the following, we let $\boldsymbol{Z}(i) \triangleq \boldsymbol{X}_d(i)$ for all i.

The asymptotic stationarity implied by the geometric ergodicity of the autoregressive process also simplifies the extensions of Lemma 2 and hence Theorem 2 in Chapter 1 to deal with the dependence between the prediction (or evaluation) point $\boldsymbol{Z} = \boldsymbol{X}_d(n)$ and the training set $T_n = \boldsymbol{Z}_n$ that occurs in the prediction of the Markovian autoregressive model (3.1). This dependence arises because after training, the last training datum $\boldsymbol{Z}(n) = \boldsymbol{X}_d(n)$ becomes the prediction input for the next process value $X(n+1)$, that is, the prediction is $\tilde{X}(n+1) \triangleq \tilde{f}_n(\boldsymbol{X}_d(n))$. Here we indicate the necessary changes to Lemma 2 and Theorem 2 in this situation.

To maintain the notation of Lemma 2, let p and p_j be the marginal densities (with respect to Lebesgue measure) of $\boldsymbol{Z}$ and $\boldsymbol{Z}(j)$ with supports S and S_j, respectively. Then (1.39) may be equivalently written as

$$\begin{aligned} q_j(\epsilon) &\triangleq P_{\boldsymbol{Z},\boldsymbol{Z}(j)}(A_{n,j}) \\ &= \int_{\boldsymbol{x} \in S_j} P_{\boldsymbol{Z}}\{z \in S: \|z - \boldsymbol{x}\| > \epsilon\} p_j(\boldsymbol{x})\, d\boldsymbol{x} \\ &= \int_{\boldsymbol{x} \in S_j \Delta S} P_{\boldsymbol{Z}}\{z \in S: \|z - \boldsymbol{x}\| > \epsilon\} p_j(\boldsymbol{x})\, d\boldsymbol{x} \\ &\quad + \int_{\boldsymbol{x} \in S_j \cap S} P_{\boldsymbol{Z}}\{z \in S: \|z - \boldsymbol{x}\| > \epsilon\} p_j(\boldsymbol{x})\, d\boldsymbol{x} \end{aligned} \tag{3.8}$$

By the geometric ergodicity condition (3.6), the first integral vanishes for j sufficiently large [since $P_{\boldsymbol{Z}(j)}(z(j) \in S_j \Delta S) \overset{j\to\infty}{\to} 0$ by the triangle-inequality $\|P_{\boldsymbol{Z}} - P_{\boldsymbol{Z}(j)}\|_V \leq \|P_{\boldsymbol{Z}} - \pi\|_V + \|\pi - P_{\boldsymbol{Z}(j)}\|_V$, where π is the stationary measure for $\{\boldsymbol{Z}(i)\}$]. The second integral can be expressed as

$$\int_{\boldsymbol{x} \in S_j \cap S} P_{\boldsymbol{Z}}\{\boldsymbol{z} \in S: \|\boldsymbol{z} - \boldsymbol{x}\| > \epsilon\} p_j(\boldsymbol{x})\, d\boldsymbol{x}$$

$$= \int_{\boldsymbol{x} \in S_j \cap S} (1 - P_{\boldsymbol{Z}}\{\boldsymbol{z} \in S: \|\boldsymbol{z} - \boldsymbol{x}\| < \epsilon\}) p_j(\boldsymbol{x})\, d\boldsymbol{x}$$

$$= P_{\boldsymbol{Z}(j)}(\boldsymbol{z}(j) \in S_j \cap S) - \int_{\boldsymbol{x} \in S_j \cap S} P_{\boldsymbol{Z}}\{\boldsymbol{z} \in S: \|\boldsymbol{z} - \boldsymbol{x}\| < \epsilon\} p_j(\boldsymbol{x})\, d\boldsymbol{x} \quad (3.9)$$

in which the first term is no greater than unity while the strict positivity of the second term given $\epsilon > 0$ follows from that of the integrand, as argued in the conclusion of the proof for Lemma 2. The desired convergence in probability then follows. Except in the application of Lemma 2 and the associated definition of $q(\epsilon)$, Theorem 2 does not assume the independence of $\boldsymbol{Z}$ from T_n and, accordingly, the changes required to have it hold for the prediction of the Markovian autoregressive model (3.1) are relatively minor:

- Condition (1.41) on the common marginal density p for $\{\boldsymbol{Z}(i)\}$ is replaced with

$$\sup_{j \in \mathbb{N}} \sup_{\boldsymbol{z} \in \mathbb{R}^d} |p_j(\boldsymbol{z})| < L_p \quad (3.10)$$

  [where p_j is the marginal density for $\boldsymbol{Z}(j)$]. This modified condition is equivalent to the original condition when the density p_π for the stationary measure π (assumed absolutely continuous with respect to Lebesgue measure) is bounded since (3.6) implies the a.e. pointwise convergence of p_j to p_π as $j \to \infty$ (by choosing B to be a point set). Thus p_j can be bounded by either the bound on p_π, when $j > N$ for some $N \in \mathbb{N}$, or by $\sup_{j=1,2,\ldots,N} \sup_{\boldsymbol{z} \in \mathbb{R}^d} |p_j(\boldsymbol{z})|$ for $1 \leq j \leq N$.
- Condition (1.46) on n is replaced with

$$\text{for } n \text{ such that } \frac{\log\left(\dfrac{\delta}{L_v}\right)}{\sum_{j=1}^{n} \log q_j(\epsilon)} < 1/2 \quad (3.11)$$

- Condition (1.50) on $q(\epsilon)$ is replaced with

$$q_i(\epsilon) = 1 - P_{\boldsymbol{Z},\boldsymbol{Z}(i)}(B_{n,i}(\epsilon)) \geq 1 - L_p(2\epsilon)^d \qquad i = 1, 2, \ldots, n \tag{3.12}$$

- In (1.53), replace $q^{n/2}(\epsilon)$ with $\prod_{j=1}^{n} q_j^{1/2}(\epsilon)$.

With these amendments, Theorem 2 follows as before.

3.6 RECURSIVE UPDATING FOR REGULARIZED RBFN PREDICTORS

As there is no substantial difficulty in doing so, we shall, where possible, develop the subsequent algorithms for a general pair of input–output processes $\{\boldsymbol{Z}(i), Y(i)\}$ rather than specifically for the autoregressive case $Y(i) \triangleq X(i)$ and $\boldsymbol{Z}(i) \triangleq \boldsymbol{X}_p(i-1)$. Thus far, both the NWRE and regularized RBFN assume that the process to be predicted admits a time-invariant regression function; in practice, as our speech prediction experiment will show, this condition does not always hold. If the regression function f drifts slowly with time index i as f_i, that is, exhibits a form of *local stationarity*, the idea of updating the regression function parameters periodically, say every l time steps, as new data arrive is intuitively appealing, particularly when it can be performed *efficiently* in a *recursive* fashion. The basis of comparison will be the standard adaptive linear estimation procedures such as the RLS algorithm. Let us consider the limiting case $l = 1$ and assume for now that n, the size of the training set and hence the number of basis functions in the estimate for f_i, is fixed. Before continuing, let us set the notations for the following discussion:

Subscripts: For *vector* and (square) *matrix quantities*, the *first subscript* refers to its *dimension*, while for a *scalar quantity*, it refers to the dimension of the associated vector or matrix quantity being indexed. The *second subscript*, if present, refers to *either* the *time index* of the training set from which the quantity is constructed (in the case of a *scalar* or *vector function*) or a *particular element* of that quantity (in the case of an *ordinary vector*). If a vector quantity's second subscript consists of the notation $a:b$, then we are referring to the subvector formed from the ath element to the bth element, inclusive.

Parenthesized Arguments: for *nonfunctional quantities*, a parenthesized argument indicates *time dependence*, that is, $\bullet(i)$ means quantity $\bullet$ uses data up to and including time step i. For functions, it indicates the usual argument.

As an example, $t_n(i) \triangleq \{(z(j), y(j))\}_{j=i-n+1}^{i}$ denotes the realized training set for the network at time step i, where in the NLAR case, this training set is formed from the time series segment $\{x(j)\}_{j=i-n-p+1}^{i}$. Then $\boldsymbol{g}_{n,i}(\bullet)$ corresponds to $\boldsymbol{g}(\bullet)$ in Eq. (42) in the Introduction and $\boldsymbol{w}_{n,j}(i)$ corresponds to the jth element of $\boldsymbol{w}$ in Eq. (15) in the Introduction when $t_n(i)$ is used in place of t_n.

Given $t_n(i)$, a realized set of input–output examples for f_i, and $\tilde{f}_{n,i}$ the corresponding regression function estimate, the problem is to recursively compute $\tilde{f}_{n,i+1}$, the estimate associated with $t_n(i+1)$, from $\tilde{f}_{n,i}$. For the NWRE, this network updating and subsequent prediction are simple, as shown in Table 3.1. If we are using some data-based method of selecting the bandwidth, it may also be advantageous to adjust the bandwidth from $h_n = h_n(i)$ to $h_n(i+1)$ at the same time. The basic order of the updating, excluding the cost of computing an updated bandwidth parameter, for the NWRE is $\mathcal{O}(1)$ and that of computing the prediction $\tilde{y}(i+1)$ is $\mathcal{O}(n)$.

For the regularized RBFN, we shall analyze the effect of the one-step updating in two stages and thereby find interesting parallels to the standard RLS estimation algorithm. In the first stage, we allow the size of the RBFN to grow with incoming data so that one weight is added per update, leading to an *augmented* network with *infinite memory* [cf. for linear adaptive filters, this growth is usually called *order recursion*; for example,

Table 3.1 NWRE Basic Fixed-Size Prediction Update Algorithm

Initialization: Assume the NWRE has been generated from $t_n(i)$ in the usual way, that is, via Eq. (30) in the Introduction with $t_n(i)$ in place of T_n.
Updating: when the new datum $(z(i+1), y(i+1))$ becomes available,

1. Replace the basis function $K(\|\bullet - z(i-n+1)\|/h_n)$ with $K(\|\bullet - z(i+1)\|/h_n)$ in Eq. (30) in the Introduction.
2. Replace the corresponding prediction target $y(i-n+1)$ with $y(i+1)$ in Eq. (30) in the Introduction.

Prediction: For the NLAR case $y(i) \triangleq x(i)$ and $z(i) \triangleq \boldsymbol{x}_p(i-1)$, set $\tilde{y}(i+2) = \tilde{f}_{n,i+1}(\boldsymbol{x}_p(i+1))$.
Iteration: $i \to i+1$ and repeat from updating step.

see Chapter 15 of Haykin (1996)]. The second stage is to simultaneously add one (new) weight and truncate the oldest weight per update, leading to a network of *fixed size* with *finite memory.*

This idea of augmenting an RBFN with incoming data was previously introduced in Platt (1991) and later Kadirkamanathan and Niranjan (1993). Compared with the latter work, our approach is developed as an optimal recursive solution to a local interpolation problem and is thus solidly grounded in the theory of recursive least-squares fitting (RLSF) which deals with noise in a principled and explicit fashion. In contrast, the sequential function estimation (SFE) approach of the latter work assumes that the training data are noise free, which may not be realistic in many applications. To ameliorate the influence of noise and to limit the network growth with their SFE approach, the latter work then proposes a growth criterion based on Hilbert function space geometry according to both prediction error and distance criteria. While such criteria may be intuitively appealing, no theoretical guidance is provided on the proper selection of the criteria parameters, nor are the conditions required for their effective application characterized. By building upon the significant body of knowledge surrounding RLSF and KRE for time-series estimation, we are able to provide analyses of our algorithmic choices and their effect on prediction performance.

3.6.1 Augmented (Infinite Memory) Case

We begin by decomposing the $(n+1) \times (n+1)$ regularized SI equation for the *combined* realized training set $t_{n+1}(i+1) = t_n(i) \cup t_n(i+1)$ as

$$\left(\begin{bmatrix} \mathbf{G}_n(i) & \boldsymbol{\gamma}_n(i+1) \\ \boldsymbol{\gamma}_n^\top(i+1) & K(0) \end{bmatrix} + \begin{bmatrix} \boldsymbol{\Lambda}_n(i) & \mathbf{0} \\ \mathbf{0}^\top & \lambda_{n+1}(i+1) \end{bmatrix} \right) \cdot \left(\begin{bmatrix} \mathbf{w}_n(i) \\ 0 \end{bmatrix} + \begin{bmatrix} \Delta\mathbf{w}_n(i) \\ w_{n+1,n+1}(i+1) \end{bmatrix} \right) = \begin{bmatrix} \mathbf{y}_n(i) \\ y(i+1) \end{bmatrix} \quad (3.13)$$

which we may write more compactly as

$$\begin{bmatrix} \mathbf{F}_n(i) & \boldsymbol{\gamma}_n(i+1) \\ \boldsymbol{\gamma}_n^\top(i+1) & K(0) + \lambda_{n+1}(i+1) \end{bmatrix} \left(\begin{bmatrix} \mathbf{w}_n(i) \\ 0 \end{bmatrix} + \begin{bmatrix} \Delta\mathbf{w}_n(i) \\ w_{n+1,n+1}(i+1) \end{bmatrix} \right) = \begin{bmatrix} \mathbf{y}_n(i) \\ y(i+1) \end{bmatrix}$$

$$\mathbf{F}_{n+1}(i+1) \cdot \mathbf{w}_{n+1}(i+1) = \mathbf{y}_{n+1}(i+1) \quad (3.14)$$

where $\boldsymbol{F}_n(i) \triangleq \boldsymbol{G}_n(i) + \boldsymbol{\Lambda}_n(i)$ and $\boldsymbol{\gamma}_n(i)$ is the vector formed from the first n elements of the last column of $\boldsymbol{G}_{n+1}(i)$, that is, $\boldsymbol{\gamma}_n(i) \triangleq [g_{n,i}(z(i))]_{1:n}$ (the notation $i{:}j$ means indices i to j inclusive). Here, as a slight generalization, $\boldsymbol{\Lambda}_n(i) \triangleq \text{diag}[\lambda_n(i-j), j = n-1, n-2, \ldots, 0]$ is the diagonal weighting matrix formed from the most recent n regularization parameters up to and including time step i. Let $\boldsymbol{w}_n(i)$ be the previously computed solution to the regularized SI equation $[\boldsymbol{G}_n(i) + \boldsymbol{\Lambda}_n(i)]\boldsymbol{w}_n(i) = \boldsymbol{y}_n(i)$ over $t_n(i)$. We assume that the new regularization parameter $\lambda_{n+1}(i+1)$ has been chosen on the basis of $t_{n+1}(i+1)$. The objective is to find the new weight $w_{n+1,n+1}(i+1)$ and the weight change vector $\Delta\boldsymbol{w}_n(i)$ to be applied to $\boldsymbol{w}_n(i)$, such that the *augmented* regularized SI equation (3.13) is satisfied. The solution is

$$w_{n+1,n+1}(i+1) = \frac{y(i+1) - (\boldsymbol{w}_n(i) + \Delta\boldsymbol{w}_n(i))^\top \boldsymbol{\gamma}_n(i+1)}{\lambda_{n+1}(i+1) + K(0)} \tag{3.15}$$

$$\Delta\boldsymbol{w}_n(i) = -\left(\boldsymbol{F}_n(i) - \frac{\boldsymbol{\gamma}_n(i+1)\boldsymbol{\gamma}_n^\top(i+1)}{\lambda_{n+1}(i+1) + K(0)}\right)^{-1} \cdot \frac{\boldsymbol{\gamma}_n(i+1)}{\lambda_{n+1}(i+1) + K(0)}(y(i+1) - \boldsymbol{w}_n^\top(i)\boldsymbol{\gamma}_n(i+1)) \tag{3.16}$$

The resultant prediction update algorithm is listed in Table 3.2.

Because $\boldsymbol{\gamma}_n(i+1)$ is also the vector of basis function outputs of the previous network from time step i in response to the newly available input $z(i+1)$, we see that the new weight $w_{n+1,n+1}(i+1)$ is merely a scaled version of the *a posteriori* estimation error, that is, the estimation error that would have been obtained had the previous weight vector $\boldsymbol{w}_n(i)$ been updated to $\boldsymbol{w}_n(i) + \Delta\boldsymbol{w}_n(i)$. In contrast, the weight change vector $\Delta\boldsymbol{w}_n(i)$ is proportional to the *a priori* estimation error, that is, the actual estimation error using the previous weight vector $\boldsymbol{w}_n(i)$ prior to any updating, similar to what occurs in the RLS algorithm. This partitioning of roles between $w_{n+1,n+1}(i+1)$ and $\Delta\boldsymbol{w}_n(i)$ is intuitively satisfying: The change $\Delta\boldsymbol{w}_n(i)$ applied to the existing weight vector attempts to account for estimation error incurred by the existing (nonupdated) network, while the new weight element $w_{n+1,n+1}(i+1)$ attempts to account for the estimation error remaining after the existing network has been updated. Analogous to the RLS algorithm, we may also expect the ratio of the MS *a priori* and the MS *a posteriori* estimation errors to converge to unity as $n \to \infty$ if the regression function being estimated is not significantly time varying. If the ratio is nonconvergent, it may be an indication that old training samples

Table 3.2 Regularized RBFN Augmented Prediction Update Algorithm

Initialization: Assume the regularized RBFN has been generated from $t_n(i)$ in the usual way, that is, via Eqs. (1) and (15) in the Introduction with $t_n(i)$ in place of T_n, and assume that $\boldsymbol{F}_n^{-1}(i)$ is known.
Updating: when the new datum $(z(i+1), y(i+1))$ becomes available,

1. Select the new regularization parameter $\lambda_{n+1}(i+1)$ and the norm-weighting matrix $\boldsymbol{U}_{n+1}(i+1)$, typically from $t_{n+1}(i+1)$.
2. Compute the new basis function vector $\boldsymbol{\gamma}_n(i+1)$.
3. Compute $\left(\boldsymbol{F}_n(i) - \dfrac{\boldsymbol{\gamma}_n(i+1)\boldsymbol{\gamma}_n^{\top}(i+1)}{\lambda_{n+1}(i+1)+K(0)}\right)^{-1}$. Note the complexity of this calculation may be reduced to $\mathcal{O}(n^2)$ if $\boldsymbol{F}_n^{-1}(i)$ is optionally propagated from time step to time step as indicated below, since the Sherman–Woodbury–Morrison formula (Hager, 1989) [or *matrix inversion lemma* in the statistical signal processing field (Haykin, 1996)] for the inverse of the sum of a given matrix and a low rank perturbation may be applied.
4. Compute the weight change vector $\Delta\boldsymbol{w}_n(i)$ according to (3.16).
5. Add the weight change vector $\Delta\boldsymbol{w}_n(i)$ to the existing network weight vector.
6. Compute the new weight $w_{n+1,n+1}(i+1)$ via (3.15).
7. Add the new basis function $K(\|\bullet - z(i+1)\|_{U_{n+1}(i+1)})$ with weight $w_{n+1,n+1}(i+1)$ to the network
8. (optional) Compute $\boldsymbol{F}_{n+1}^{-1}(i+1)$ from $\boldsymbol{F}_n^{-1}(i)$ with complexity $\mathcal{O}(n^2)$ via a partitioned matrix inverse formula applied to the decomposition (3.14).

Prediction: For the NLAR case $y(i) \triangleq x(i)$ and $z(i) \triangleq \boldsymbol{x}_d(i-1)$, set $\tilde{x}(i+2) = \tilde{f}_{n+1,i+1}(\boldsymbol{x}_d(i+1))$.
Iteration: $i \to i+1$, $n \to n+1$ and repeat from *Updating step*.

are no longer representative of the regression function behavior currently being estimated. For this situation, the effective *memory* of the RBFN can be limited by fixing its size to n weights/basis functions computed from the most recent n training data available, which leads us to the second stage of updating described next.

3.6.2 Fixed-Size (Finite Memory) Case

Let us return to the original task and assume that the size of the RBFN is fixed at n weights/basis functions. The desire is to relate $\boldsymbol{w}_n(i+1)$, the weights satisfying the regularized SI equation over $t_n(i+1)$, to the previously computed weights $\mathbf{w}_n(i)$, which do the same for $t_n(i)$. Before we do so, let us establish the notations. Decompose the $n \times n$ regularized SI equation for the previous training set $t_n(i)$ as

$$\begin{bmatrix} \lambda_n(i-n+1)+K(0) & \boldsymbol{\beta}_{n-1}^{\top}(i) \\ \boldsymbol{\beta}_{n-1}(i) & \mathbf{F}_{n-1}(i) \end{bmatrix} \begin{bmatrix} w_{n,1}(i) \\ \mathbf{w}_{n,2:n}(i) \end{bmatrix} = \begin{bmatrix} y(i-n+1) \\ \mathbf{y}_{n,2:n}(i) \end{bmatrix}$$

$$\mathbf{F}_n(i) \cdot \mathbf{w}_n(i) = \mathbf{y}_n(i) \quad (3.17)$$

where $\boldsymbol{\beta}_{n-1}(i)$ is the vector of the last $n-1$ elements of the first column of the previous interpolation matrix $\mathbf{G}_n(i)$, that is, $\boldsymbol{\beta}_{n-1}(i) \triangleq [g_{n,i}(z(i-n+1))]_{2:n}$. This time, the objective is to find $\Delta\mathbf{w}_{n,2:n}(i)$ and $w_{n,n}(i+1)$ satisfying

$$\begin{bmatrix} \mathbf{F}_{n-1}(i) & \boldsymbol{\gamma}_{n-1}(i+1) \\ \boldsymbol{\gamma}_{n-1}^{\top}(i+1) & \lambda_n(i+1)+K(0) \end{bmatrix} \left(\begin{bmatrix} \mathbf{w}_{n,2:n}(i) \\ 0 \end{bmatrix} + \begin{bmatrix} \Delta\mathbf{w}_{n,2:n}(i) \\ w_{n,n}(i+1) \end{bmatrix} \right) = \begin{bmatrix} \mathbf{y}_{n,2:n}(i) \\ y(i+1) \end{bmatrix}$$

$$\mathbf{F}_n(i+1) \cdot \mathbf{w}_n(i+1) = \mathbf{y}_n(i+1) \quad (3.18)$$

In other words, the new weight vector for the updated network can be considered the result of: (i) shifting the last $n-1$ weights in the old weight vector $\mathbf{w}_n(i)$, which are associated with the most recent $n-1$ data in $t_n(i)$, upward into positions 1 to $n-1$ and setting the nth element to zero; (ii) adding a perturbation $\Delta\mathbf{w}_{n,2:n}(i)$ to the shifted vector; and (iii) adding a new weight $w_{n,n}(i+1)$ in the nth position. It is not difficult to show that the resultant update equations become

$$w_{n,n}(i+1) = \frac{y(i+1) - (\mathbf{w}_{n,2:n}(i) + \Delta\mathbf{w}_{n,2:n}(i))^{\top}\boldsymbol{\gamma}_{n-1}(i+1)}{\lambda_n(i+1)+K(0)} \quad (3.19)$$

$$\Delta\mathbf{w}_{n,2:n}(i) = \left(\mathbf{F}_{n-1}(i) - \frac{\boldsymbol{\gamma}_{n-1}(i+1)\boldsymbol{\gamma}_{n-1}^{\top}(i+1)}{\lambda_n(i+1)+K(0)} \right)^{-1} \cdot \left[w_{n,1}(i)\boldsymbol{\beta}_{n-1}(i) - \frac{\boldsymbol{\gamma}_{n-1}(i+1)}{\lambda_n(i+1)+K(0)} \cdot (y(i+1) - \mathbf{w}_{n,2:n}^{\top}(i)\boldsymbol{\gamma}_{n-1}(i+1)) \right] \quad (3.20)$$

Except for the additional term $w_{n,1}(i)\boldsymbol{\beta}_{n-1}(i)$ in (3.20), the forms of the update equations for this fixed-size case are identical to those for the augmented case. The additional term can be regarded as embodying the effect of weight vector augmentation from size n to $n+1$, followed by truncation to the weights computed from the most recent n training data. We summarize the prediction update algorithm for the fixed-size case in Table 3.3. Note that formula (3.22) in *Updating* step 9 follows from the identity

$$\begin{bmatrix} 1 & \mathbf{0}^{\top} \\ \mathbf{0} & \boldsymbol{F}_{n-1}^{-1}(i+1) \end{bmatrix} = \boldsymbol{F}_n^{-1}(i+1)\begin{bmatrix} \lambda_n(i-n+2)+K(0) & \boldsymbol{\beta}_n^{\top}(i+1)\boldsymbol{F}_{n-1}^{-1}(i+1) \\ \boldsymbol{\beta}_n(i+1) & \boldsymbol{I} \end{bmatrix} \tag{3.21}$$

Although the parallels between the recursive update algorithms described here and those in the RLS algorithm are interesting in their own right, one must be careful not to conclude that the algorithms presented are merely expressions of the RLS algorithm after a nonlinear mapping $z(i) \in \mathbb{R}^p \mapsto \boldsymbol{g}_{n,i-1}(z(i)) \in \mathbb{R}^n$. We can see this difference clearly in the fact that infinite-memory regularized RBFNs require an infinite number of weights/basis functions; fixed-size regularized RBFNs can only have a finite memory of the same size. This condition stands in contrast to the situation with the RLS filter where a fixed number of weights are updated to reflect all the past history of the input data. Of course, the exponentially weighted variant of the RLS algorithm is commonly used in practice and one can argue that its memory is, for all practical purposes, limited. Indeed, the introduction of the exponentially weighted variant of the RLS algorithm was motivated by the heuristic that decaying memory would improve estimation when the input–output processes are nonstationary, although it has now been established that this notion is, in fact, generally incorrect Haykin et al. (1997a). In this respect, the fixed-size regularized RBFN is somewhat more explicit in the way it deals with nonstationarity.

With both the augmented and fixed-size update algorithms, their computational efficiency is derived from the low rank of the perturbation applied to the existing interpolation matrix at a given time step through augmentation and addition, respectively. Exploiting the matrix inversion lemma can then reduce the update complexity to $\mathcal{O}(n^2)$ (for n basis functions) per time step. As may be expected, the experimental results for speech prediction show that these *partial update* algorithms can result in

Table 3.3 Regularized RBFN Fixed-Size Prediction Update Algorithm

Initialization: assume the regularized RBFN has been generated from $t_n(i)$ in the usual way, that is, via Eqs. (1) and (15) in the Introduction with $t_n(i)$ in place of T_n, and assume that $\boldsymbol{F}_{n-1}^{-1}(i)$ is known.
Updating: when the new datum $(z(i+1), y(i+1))$ becomes available,

1. Select the new regularization parameter $\lambda_n(i+1)$ and the norm weighting matrix $\boldsymbol{U}_n(i+1)$, typically from $t_n(i+1)$.
2. Compute the new basis function vector $\boldsymbol{\gamma}_{n-1}(i+1)$.
3. Compute $\left(\boldsymbol{F}_{n-1}(i) - \dfrac{\boldsymbol{\gamma}_{n-1}(i+1)\boldsymbol{\gamma}_{n-1}^{\mathsf{T}}(i+1)}{\lambda_{n+1}(i+1)+K(0)}\right)^{-1}$. Complexity can be reduced to $\mathcal{O}(n^2)$ if $\boldsymbol{F}_{n-1}^{-1}(i)$ is optionally propagated from time step to time step as in step 3 of the updating procedure in Table 3.2.
4. Compute the shifted weight change vector $\Delta\boldsymbol{w}_{n,2:n}(i)$ according to (3.20).
5. Compute the new weight $w_{n,n}(i+1)$ via (3.19).
6. Delete the basis function $K(\|\bullet - z(i-n+1)\|)_{U_n(i)})$ and its weight $w_{n,1}$ associated with the oldest data in $t_n(i)$ from the network.
7. Add the shifted weight change vector $\Delta\boldsymbol{w}_n(i)$ to the remaining $n-1$ network weights.
8. Add the new basis function $K(\|\bullet - z(i+1)\|_{U_n(i+1)})$ with weight $w_{n,n}(i+1)$ to the network
9. (optional) Compute $\boldsymbol{F}_n^{-1}(i+1)$ from $\boldsymbol{F}_{n-1}^{-1}(i)$ with complexity $\mathcal{O}(n^2)$ via a partitioned matrix inverse formula applied to the decomposition (3.17). Hence compute $\boldsymbol{F}_{n-1}^{-1}(i+1)$ with complexity $\mathcal{O}(n^2)$ via
$$\boldsymbol{F}_{n-1}^{-1}(i+1) = (\boldsymbol{I} - \boldsymbol{h}_{n-1}(i+1)\boldsymbol{\beta}_{n-1}^{\mathsf{T}}(i+1))^{-1}\boldsymbol{H}_{n-1}(i+1) \tag{3.22}$$
where $\boldsymbol{h}_{n-1}(i+1)$ is the vector formed from the last $n-1$ elements of the first column of $\boldsymbol{F}_n^{-1}(i+1)$ and $\boldsymbol{H}_{n-1}(i+1)$ is the lower right $(n-1)\times(n-1)$ submatrix of $\boldsymbol{F}_n^{-1}(i+1)$.

Prediction: for the NLAR case $y(i) \triangleq x(i)$ and $z(i) \triangleq \boldsymbol{x}_d(i-1)$, set $\tilde{x}(i+2) = \tilde{f}_{n,i+1}(\boldsymbol{x}_d(i+1))$.
Iteration: $i \to i+1$ and repeat from *Updating step*.

loss of tracking and degraded performance compared to a *full update* algorithm in which the bandwidth and/or regularization parameter is updated for *all* entries of the regularized interpolation matrix $\boldsymbol{F}_n(i)$, not just those involving the new basis function vectors $\boldsymbol{\gamma}_n(i+1)$ (in the case of the augmented updates) and $\boldsymbol{\gamma}_{n-1}(i+1)$ (in the case of the fixed-size updates). The update complexity per time step in this full update case is naturally greater at $\mathcal{O}(n^3)$ compared to the partial update case. Nevertheless, the recursive update algorithms for both cases provide useful insight into the essential character and operation of the dynamic regularized RBFN as a time-series estimator.

3.7 APPLICATION TO SPEECH PREDICTION

For a benchmark problem with real-world data, we turn to speech prediction. That the human speech signal is generally nonlinear and nonstationary is well known; even so, the linear prediction of speech with *analytic* methods, such as the least mean-squares (LMS), RLS, and Kalman algorithms (Haykin 1996), and *synthetic* methods such as CELP (Schroeder and Atal, 1985) has been met with surprising success. Of course, these results are achieved after significant prior knowledge regarding the characteristics of human speech has been carefully embedded into the corresponding methods to realize maximum performance. In contrast, we should emphasize that our interest in speech as the test signal for the proposed algorithms is limited to the characterization of the gains possible from nonlinear and nonstationary processing and should not be taken to imply that the proposed predictors (in their current form) are either practical or optimally tuned for actual speech prediction applications such as speech coding. Further speech-specific research and evaluation would clearly be necessary to reach that state. That said, the results of the following experiments in which both the partial and full update algorithms for the fixed-size network case are evaluated (albeit with different motivations) do offer evidence of the performance gains possible when the nonlinearity and nonstationarity of speech signals are addressed.

3.7.1 Experiment 1: Partial Update Algorithm for Fixed-Size Networks

We begin by giving some results for the fixed-size partial update algorithm of Table 3.3. At this stage of development, we focus our attention on the practical issues of predictor tracking stability and performance versus the fixed-size full update algorithm.

Description of Speech Data We use a 10,000-point speech sample of a male voice recorded at 8 kHz and 8 bits per sample while speaking the sentence fragment "When recording audio data." The speech data, which appear to have no discernible noise, are approximately zero mean and normalized to unit total amplitude range. Applying the Mann–Whitney rank-sum test as described in Section 3.7.2 rejects the null hypothesis that the speech sample is that of a stationary linear process with a maximum sample Z statistic of less than -13 (a Z statistic of less than -3 is considered grounds for strong rejection), hence indicating a high probability of nonlinearity in the speech sample.

Approach Using Regularized RBFNs In the main, we follow the same approach as in the full update case discussed in Section 3.7.2 except for the following modifications:

Input Order: a common input order of $p = 50$ is used for each network (unless otherwise indicated).

Regularization Parameter: for a given network, fixed for the duration of prediction over the input signal, that is, $\lambda_n(i+1) = \lambda_n(i)$ for all i.

Update Algorithm: except during reset (see following), we follow Table 3.3, where the updated norm-weighting matrix $\boldsymbol{U}_n(i)$ is computed according to the input data covariance formula described in the corresponding section below for the full update case. The updated norm-weighting matrix, however, is applied only to the new basis functions in the updated column $\boldsymbol{\gamma}_{n-1}(i+1)$ in (3.18) to maintain consistency with the usual SI fitting relation $\tilde{\boldsymbol{y}}_n(i+1) = \boldsymbol{G}_n(i+1)\boldsymbol{w}_n(i+1)$, where $\tilde{\boldsymbol{y}}_n(i+1)$ is the estimate of $\boldsymbol{y}_n(i+1)$ produced by the network at time step $i+1$.

Reset Algorithm: as can be seen, the partial updating algorithm implies that the networks produced no longer exactly solve the interpolation problem Eqs. (11) and (12) in the Introduction [since with partial updates the interpolation matrix $\boldsymbol{G}_n(i)$ is not identical to the one specified by interpolation problem over $t_n(i)$]. The accumulation of these partial updates to the interpolation matrix over many consecutive time steps can lead to a loss of tracking and instability. To counteract this problem, we monitor the prediction error $\epsilon_n(i+1)$ of the dynamic network at each time step i and *reset* the network, that is, restart the partial update algorithm from *Initialization* step 1 of Table 3.3, when one of two possible conditions, denoted RC.1 and RC.2, are met:

$$|\epsilon_n(i+1) - \mu(\epsilon_n(i), m)| > \kappa\sigma(\epsilon_n(i), m) \qquad \text{(RC.1)}$$

$$|\epsilon_n(i+1)| - \mu(|\epsilon_n(i)|, m) > \kappa\sigma(|\epsilon_n(i)|, m) \qquad \text{(RC.2)}$$

where for a sequence $\{a(i)\}$, $\mu(a(i), m)$ is the sample mean and $\sigma(a(i), m)$ is the sample standard deviation of $\{a(j)\}_{j=i-m+1}^{i}$. Thus a predictor reset occurs when a probable large deviation (as set by the *window* parameter m and *threshold* parameter κ) has occurred in either the two-sided (RC.1) or one-sided (RC.2) prediction error. For our experiment, we use reset condition RC.1 with window $m = n = 100$ and threshold $\kappa = 4$, as there appears to be no substantial difference in performance compared to condition RC.2. In the ideal case that prediction error is a white Gaussian process, the choice of κ corresponds to a large deviation probability of approximately 0.0063%. Not unexpectedly, the actual reset rate in the experiment is quite a bit greater due to heavy tails in the prediction error density.

These design decisions yield networks with moderate computational complexity and reasonable performance that suit the basic purpose of demonstrating the partial update algorithm for fixed-size networks. Further optimization of the design choices with their concomitant increased computational load are no doubt possible but will not be pursued here.

Dynamic Updating and Regularization for Speech Prediction

Using Figs. 3.1 and 3.2, we can briefly argue for the practical utility of dynamic updating and regularization for speech prediction.

In Fig. 3.1, we compare the initial predictions of a dynamic predictor trained according to the partial update algorithm (without reset) for a $n = 100$, $\lambda = 0.01$ fixed-size network with those from a static $n = 250$, $\lambda = 0.01$ predictor whose network parameters are frozen after the initial training. Not surprisingly, even with more than twice the number of basis functions, the static predictor quickly loses track of the speech signal in the transition from a quickly to a slowly varying portion of the input signal as shown in Fig. 3.1. The dynamic predictor, however, is able to adapt and maintain its prediction performance. Regarding regularization, although RLSF theory implies that $\lambda = 0$ is a consistent choice when no noise is present, in practice some regularization is necessary because the likelihood of a singular/ill-conditioned interpolation matrix $\boldsymbol{G}_n(i)$ increases as n increases. Empirically, this effect appears especially pronounced for small values of p in which the predictor output is more sensitive to individual inputs in the input vector. An example of this phenomenon can be seen in Fig. 3.2, where we contrast the predictions for two partially updated (without reset) fixed-sized predictors, one of which is trained with a fixed $\lambda = 0.1$ and the other with a fixed $\lambda = 0.01$. Again, it is evident

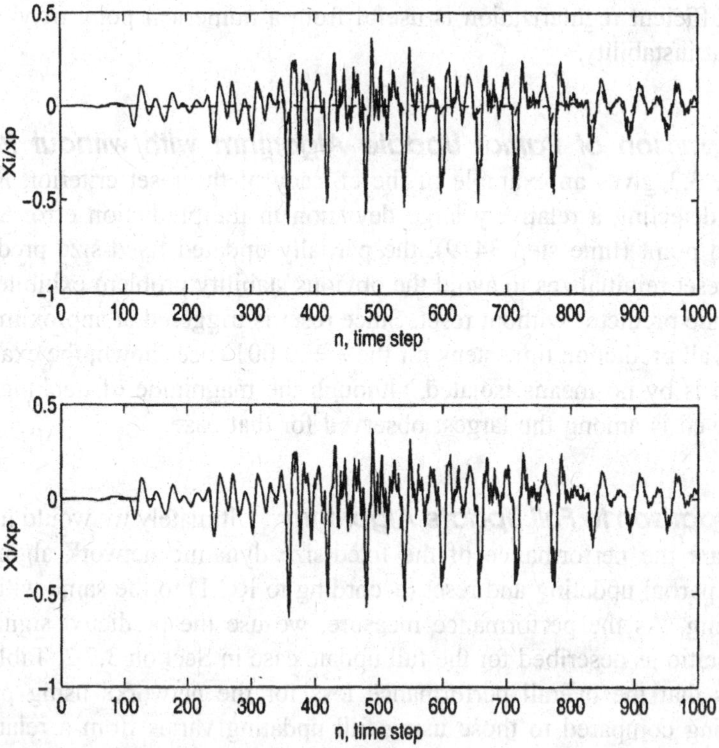

Figure 3.1 Tracking ability of static (upper) versus dynamic (lower) predictors (solid is actual; dashed is predicted).

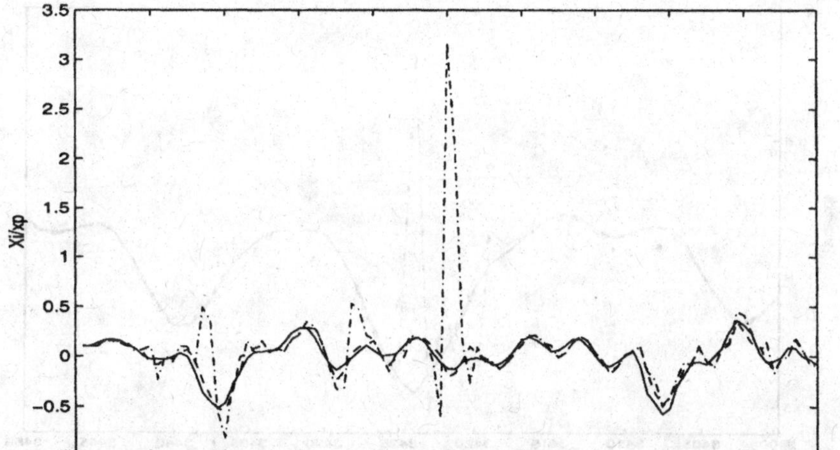

Figure 3.2 Regularized versus nonregularized predictors, $n = 100, p = 2$ (solid is actual; dashed is predicted for $\lambda = 0.1$; dash–dot is predicted for $\lambda = 0.01$).

that sufficient regularization is useful from a numerical point of view to combat instability.

Comparison of Partial Update Algorithm with/without Reset Figure 3.3 gives an example of the efficacy of the reset criterion RC.1. After detecting a relatively large deviation in the prediction error at the starred point (time step 3419), the partially updated fixed-size predictor with reset reinitializes to avoid the obvious stability problem exhibited by the same predictor without reset. Since reset is triggered at approximately 1% of all prediction time steps for the $\lambda = 0.001$ case shown, the example shown is by no means isolated, although the magnitude of tracking loss displayed is among the largest observed for that case.

Comparison to Full Update Algorithm Ultimately we would like to compare the performance of the fixed-size dynamic network algorithm using partial updating and reset (according to RC.1) to the same with full updating. As the performance measure, we use the predicted signal-to-noise ratio as described for the full update case in Section 3.7.2. Table 3.4 shows that the overall performance loss for the networks using partial updating compared to those using full updating varies from a relatively minimal 0.28 dB for the $\lambda = 0.00001$ predictor to a more substantial 0.74 dB for the $\lambda = 0.0001$ predictor.

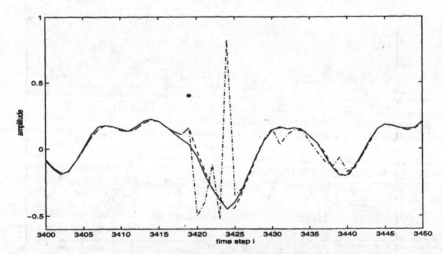

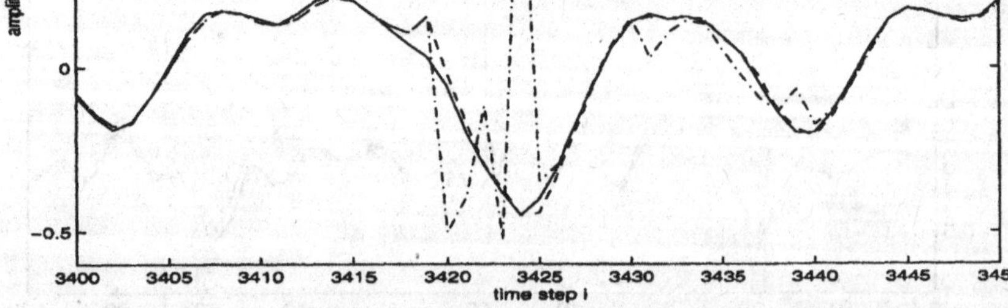

Figure 3.3 Partial update algorithm, fixed-size case with reset versus without reset (solid is actual; dashed is predicted with reset; dash–dot is predicted without reset; star indicates reset point).

Table 3.4 Prediction Performance of Fixed-Size Algorithm with Partial Update versus Full Update

RBF Reg. Parameter	% of Pred. Resets	PSNR (dB)		
		Partial Update	Full Update	Partial − Full
$\lambda = 0.001$	0.94	14.37	14.73	−0.36
$\lambda = 0.0001$	0.92	13.71	14.45	−0.74
$\lambda = 0.00001$	0.87	13.91	14.19	−0.28

From a computational standpoint, the figures for the percentage of points at which reset is triggered indicate that the partially updated fixed-size dynamic predictor has only 1% of the computational complexity of the corresponding fully updated fixed-size dynamic predictor. While this reset rate is two orders of magnitude larger than expected in the ideal case of white Gaussian prediction errors, it still easily satisfies the basic distribution-free upper bound of $\frac{1}{16} = 6.25\%$ implied by the Chebyshev inequality, namely

$$\text{Prob}\{|\epsilon_n(i+1) - \mu^*(\epsilon_n(i))| > \kappa\sigma^*(\epsilon_n(i))\} < \frac{1}{\kappa^2}$$

where $\mu^*(\bullet)$ and $\sigma^*(\bullet)$ are the true, that is, distributional, mean and standard deviation of a process $\bullet$.

Figure 3.4 shows the points of reset for the $\lambda = 0.001$ partially updated fixed-size dynamic predictor from time steps 2500 to 5000. It is interesting to note how in this segment the predictor resets occur at points of a regime shift within the speech sample. Because the performance/computational trade-off between the two update techniques is influenced by several factors such as the length of prediction, the speech segment being predicted, and so forth, further characterization is necessary to make more definitive statements; nonetheless, the results can be considered encouraging.

Comparison to Previous Work The same speech signal was also used as part of two previous studies, both of which are based on *pipelined recurrent neural networks (PRNN)* followed by standard linear adaptive filters (Haykin and Li, 1995; Baltersee and Chambers, 1998; Mandic and Baltersee, 1999). PRNNs represent another method of modeling nonsta-

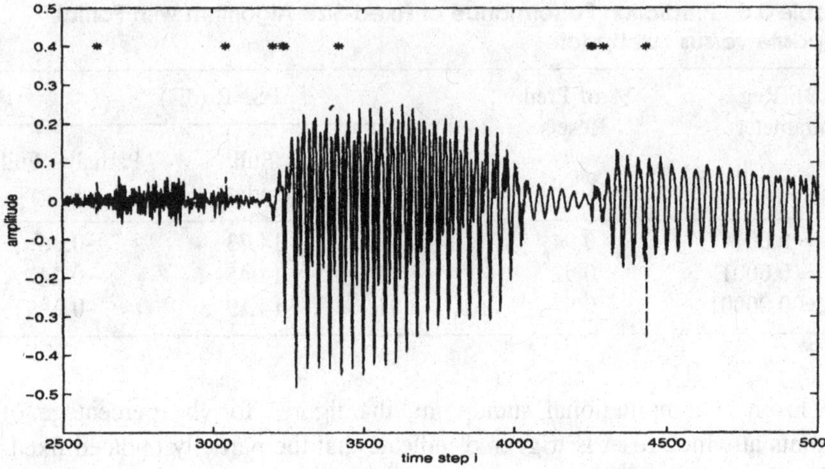

Figure 3.4 Reset points in second 2500 predicted points for partial update algorithm, fixed-size case (solid is actual; dashed is predicted; stars indicate reset positions).

tionary dynamics based on the use of explicit feedback between modular network elements, each of which is itself a (recurrent) neural network. While considerably different in the details of their architectures and training methods, they do share the common principle of continuously adapting their network parameters to minimize their squared prediction error and thus track nonstationary signal characteristics. Comparing our results in Table 3.4 with those of Table I in Baltersee and Chambers (1998), we see that even our worst-performing case of partial updating yields a PSNR of 13.71 dB which is 0.12 dB better than the best PSNR of 13.59 dB listed in Table II for a hybrid extended RLS (ERLS)-trained PRNN followed by a 12th-order RLS filter. To be fair, however, the $\mathcal{O}(n^2)$ computational complexity of our partial updating method with $n = 100$ centers is most likely somewhat greater than that of the ERLS PRNN with three 8-input single-neuron modules used in Baltersee and Chambers (1998). On the other hand, their predictor has the benefit of an additional level of RLS linear prediction not (yet) present in our scheme, and their performance figures are reported using the variance rather than mean-squared value of the prediction error, both of which should bias their results upwards compared to our PSNR figures (see the discussion in Section 3.7.2). Of course, it would be premature to draw any substantial conclusions on the basis of a single speech signal, which leads us to consider the more comprehensive suite of longer, phonetically balanced male and female speech signals in the next experiment. Over this test suite, we shall also show the performance increase possible from employing a similar final level of RLS linear prediction.

3.7.2 Experiment 2: Full Update Algorithm for Fixed-Size Networks

As we previously mentioned, our objective in this set of experiments is to demonstrate that even without significant tuning, the dynamic regularized RBFN can provide a nontrivial improvement in prediction signal-to-noise (SNR) over the standard LMS/RLS algorithm-based predictors. We also indicate the further improvement possible in exploiting the residual correlations between the predictions of several dynamic regularized RBFNs and the predicted, that is, desired, speech signal by way of an additional stage of RLS estimation.

Description of Speech Data The speech data to be predicted consist of samples from 10 different male and 10 different female speakers, each reading a distinct phonetically balanced sentence. In their original format, the continuous speech signals were 16-bit linear pulse code modulated (PCM) and sampled at 16 kHz rate with 8 kHz bandwidth. These samples were subsequently filtered by a third-order Butterworth filter with a cutoff frequency of 3.2 kHz, decimated to 8 kHz rate, and recentered to zero mean. Both the original and final speech signals are of high quality with little discernible background noise. The sentence samples and some of their key characteristics as discrete time series are summarized in Tables 3.5 and 3.6. As can be seen from these tables, the total length of a speech signal being tested varies from approximately 2.5 to 4 s.

Before beginning, it is useful to quantify the degree of nonlinearity in the speech samples, as this factor will ultimately determine the gains possible in our approach. Using some software for chaotic time-series analysis developed by the chemical reactor engineering group at Delft University of Technology in the Netherlands (Schouten and Bleek, 1994), the method of surrogate data analysis (Theiler et al., 1992) with a Mann–Whitney rank-sum test rejects the null hypothesis that each speech sample is that of a stationary linear process with a maximum sample Z statistic of less than -13 in each case (a Z statistic of less than -3 is considered grounds for strong rejection). This result indicates that significant benefit from nonlinear processing should be possible.

Table 3.5 Male Speech Sample Parameters

ID	No. of samples	Sentence
m130	20215	Type out three lists of orders.
m131	23525	The harder he tried, the less he got done.
m132	25128	The boss ran the show with a watchful eye.
m133	25129	The cup cracked and spilled its contents.
m134	23260	Paste can cleanse the most dirty brass.
m135	29073	The slang word for raw whiskey is booze.
m136	27746	It caught its hind paw in a rusty trap.
m137	26724	The wharf could be seen at the farthest shore.
m138	22435	Feel the heat of the weak dying flame?
m139	20877	The tiny girl took off her hat.

Approach Using Regularized RBFNs The particular approach taken is to treat each speech sample as a realization of a discrete-time Markov process of order p obeying (3.1). For one-step-ahead (OSA) prediction $k = 1$, we consider the limiting case of per time step updating, that is, $l = 1$. Key design issues to consider are:

Input Order: preliminary experiments showed that, for a given speech sample, the prediction performance of the dynamic regularized RBFN varied with the order p depending on the local characteristics of the speech over which the network was operating. For example, in the transition periods between voiced, unvoiced, and silent segments,

Table 3.6 Female Speech Sample Parameters

ID	No. of samples	Sentence
f150	21530	The young kid jumped the rusty gate.
f151	25471	Guess the results from the first scores.
f152	25064	A salt pickle tastes fine with ham.
f153	24674	The just claim got the right verdict.
f154	24361	These thistles bend in a high wind.
f155	22098	Pure bred poodles have curls.
f156	26954	The tree top waved in a graceful way.
f157	32522	The spot on the blotter was made by green ink.
f158	27220	Mud was spattered on the front of his white shirt.
f159	23306	The cigar burned a hole in the desk top.

networks with small p, for example, $p = 10$, were generally found to perform better than those with large p, for example, $p = 50$. Conversely, within a given type of speech segment, the networks with larger p tended to be the better predictors. While techniques for estimating the order of NLAR processes have been recently proposed (Auestad and Tjøstheim, 1990), for computational simplicity three fixed-sized networks with $p = 10$, 30, and 50 are run in parallel for each speech sample and, as we shall see later, linearly combined.

RBFN Parameters: based on some previous work (Yee and Haykin, 1995), each of the networks is chosen to have:

Network Size: A fixed size of $n = 100$ basis functions is used. This fixed size corresponds to the assumption that a useful memory for the networks is 12.5 ms, the average length of the 5 to 20 ms window of stationarity usually associated with speech.

Basis Function: the "smooth" [in the sense of satisfying Eq. (12) in the Introduction with the choice (13) in (11)] Gaussian basis function $K(r) = \exp(-r^2/2)$ is used.

Norm-Weighting Matrix: common to all basis functions is a diagonal norm-weighting matrix $\boldsymbol{U}_n(i)$ whose inverse, at time step i, is set to p times the diagonal of the empirical covariance matrix for the input samples in $t_n(i)$. This particular form of the norm-weighting matrix allows the multidimensional network basis function to be decomposed into a p-fold product of one-dimensional (Gaussian) kernels, each with bandwidth parameter equal to the variance of a particular window over $t_n(i)$. In the one-dimensional i.i.d. density estimation setting, such a form of bandwidth has been shown to be consistent in the L_1 sense (Devroye and Györfi, 1985).

Regularization Parameter: for each of the three networks ($p = 10$, 30, and 50), the regularization parameter for each time step is selected as the value that minimizes the generalized cross-validation (GCV) criterion function evaluated over 1000 logarithmically spaced points from $\lambda_{\min}$ to $\lambda_{\max}$ for that network as given in Table 3.7. Since the speech signals are largely noise free, the upper bound on $\lambda_n(i+1)$ prevents undue overregularization while the lower bound is necessary to ensure the numerical nonsingularity of the regularized SI matrix at each time step. The slight differences in the evaluation limits account for the varying degrees of sensitivity of each network to these two conditions.

Table 3.7 GCV Criterion Function Evaluation Limits

Network no.	p	λ_{min}	λ_{max}
1	50	0.0001	0.001
2	30	0.00001	0.01
3	10	0.0001	0.01

Update Algorithm: because the norm weighting matrix $\boldsymbol{U}_n(i)$ is updated for *all* network basis functions when new data arrive, the update from $\boldsymbol{F}_n(i)$ to $\boldsymbol{F}_n(i+1)$ is full rank and hence (3.18) must be solved directly without using the recursion aids (3.19) and (3.20). It was found in previous experiments (Yee and Haykin, 1995) that the speech samples were sufficiently nonstationary so that without careful choice of the update parameters indicated in the first updating step of each algorithm, the recursively updated fixed-size network outputs would frequently loose track of the speech samples within an order of n time steps from the last full-rank update. Notwithstanding the results of the previous section, the issue of how best to select the update parameters in the recursive fixed-size update algorithm so as to minimize performance loss from partial updating remains an open question.

Comparison to Linear RLS Algorithm and Previous Work The performance measure we shall use is the *predicted signal-to-noise ratio (PSNR)* defined for an actual or input signal sequence $\{y(i)\}_{i=1}^{N}$ by

$$\text{PSNR (dB)} \triangleq 10 \log_{10}(\widetilde{\sigma}_y^2/\widetilde{\sigma}_\epsilon^2) \tag{3.23}$$

where $\widetilde{\sigma}_y^2$ and $\widetilde{\sigma}_\epsilon^2$ are the actual and error signal *powers* estimated by

$$\widetilde{\sigma}_y^2 \triangleq \frac{1}{N}\sum_{i=1}^{N} y^2(i), \qquad \widetilde{\sigma}_\epsilon^2 \triangleq \frac{1}{N}\sum_{i=1}^{N} \epsilon^2(i) \text{ where } \epsilon(i) \triangleq y(i) - \tilde{y}(i) \tag{3.24}$$

and $\tilde{y}(i)$ is the network prediction for actual signal $y(i)$. The PSNR can be considered a measure of the *generalization* performance of the dynamic network, since in our NLAR case, each prediction $\tilde{y}(i) = \tilde{x}(i+1)$ at time step i is for the first time series point *outside* the window $\{x(j)\}_{j=i-(n+p-1)}^{i}$ of data effectively seen during training [$n+p$ sequential data are needed to form $t_n(i)$]. This effective training window, along with the predicted point, shifts forward in time as the dynamic network advances through the entire input signal sequence. Although, strictly speaking, the test set per time step is a single (out-of-sample) point, by iterating the training/prediction cycle over the available input time series (the number of samples in each speech signal listed in Tables 3.5 and 3.6 less $n+p$ samples for initialization), this pointwise prediction performance can be averaged to gauge the generality of our method. For example, the PSNR figure in Table 3.8 for network 1 operating on signal m130 is computed according to (3.23) and (3.24) with $N = 20{,}215 - 100 - 50 = 20{,}065$ effective test data.

Note that in our case of zero-mean input signals, because we use estimated signal powers rather than estimated signal variances as is sometimes the case in defining the PSNR, the following performance figures are somewhat conservative, for example, a non-zero mean level of error will degrade performance by the former definition but not by the latter definition. That said, the PSNR results of the three RBFN predictors individually and jointly (as will be explained) over the complete speech samples can be found in Tables 3.8, 3.9, 3.11, and 3.12. Summary tables of minimum, average, and maximum performance gains are listed in Tables 3.10, 3.13, and 3.14 for the male only, female only, and joint male/female samples, respectively. The first four lines of each table list the individual predictor performances along with their arithmetic average. We see an average gain of 1.65 dB of the basic regularized RBFN predictors over the RLS predictor for the male speech samples while the average gain for the female speech samples is somewhat better at 2.67 dB. Over both the male and female speech samples, the average gain is 2.2 dB. The RLS predictor performance reported in the fifth line (with the corresponding autoregressive input order and exponential weight in parentheses) is the best one observed in a series of experiments for which the parameters vary, as in Table 3.15. To allow a fair assessment of the gains possible from nonlinear versus linear prediction, the maximum order p of the linear predictor is set to 50, the same as for the RBFN. With regards to nonlinear speech predictors, these figures are in general agreement with those in previously published work (Townshend, 1991; Maria and Figueiras-Vidal, 1995; Baltersee and Chambers, 1998; Mandic and Baltersee, 1999). In particular, Townshend (1991) reported an increase in prediction gain of 2.8 dB when a nonlinear predictor is trained on the residuals of a time-varying LPC predictor, which may be considered a *linear–nonlinear* processing scheme. As previously mentioned in Section 3.7.1, (Haykin

Table 3.8 Overall Experimental Results for Speech Prediction, Samples m130 to m134

Network no.	m130	m131	m132	m133	m134
1	14.02	13.76	15.30	9.93	12.79
2	14.91	13.83	15.53	9.91	12.51
3	14.04	13.49	15.24	11.27	13.40
NL avg.	14.32	13.69	15.36	10.37	12.90
RLS	11.76	11.77	14.58	10.85	11.83
(auto)	(14, 0.99)	(50, 0.999)	(50, 0.99)	(50, 0.999)	(50, 0.999)
RLS	16.07	15.09	17.08	12.36	14.75
(NL)	(1, 0.98)	(1, 0.99)	(1, 0.98)	(1, 0.99)	(1, 0.999)
RLS	16.43	15.58	17.73	13.14	15.31
(NL + auto)	(1 + 6, 0.99)	(4 + 14, 0.999)	(1 + 10, 0.99)	(3 + 6, 0.999)	(3 + 14, 0.999)

Note: All PSNR in dB.

Table 3.9 Overall Experimental Results for Speech Prediction Example, Samples m135 to m139

Network no.	m135	m136	m137	m138	m139
1	17.37	12.22	15.41	15.09	15.41
2	17.30	12.81	15.43	15.13	17.30
3	16.69	13.22	15.74	15.52	16.69
NL avg.	17.12	12.75	15.53	15.25	15.53
RLS	15.48	10.92	13.24	12.48	13.43
(auto)	(46, 0.99)	(10, 0.99)	(50, 0.99)	(50, 0.999)	(50, 0.999)
RLS	18.63	14.42	17.05	16.92	16.20
(NL)	(1, 0.99)	(1, 0.99)	(1, 0.999)	(1, 0.99)	(1, 0.999)
RLS	19.35	14.75	17.42	17.11	16.86
(NL + auto)	(1 + 6, 0.99)	(1 + 4, 0.99)	(2 + 6, 0.999)	(5 + 10, 0.999)	(1 + 14, 0.999)

Note: All PSNR in dB.

Table 3.10 Summary of Gains in Experimental Results for Speech Prediction, Male Speech Samples

Gain	Min.	Avg.	Max.
NL avg. over RLS(auto)	−0.48	1.65	2.77
RLS(NL) over NL avg.	0.67	1.58	1.99
RLS(NL + auto) over RLS(NL)	0.19	0.51	0.78
Total over RLS(auto)	2.29	3.73	4.67

Note: All figures in dB.

and Li, 1995), and (Baltersee and Chambers, 1998; Mandic and Baltersee, 1999) considered enhancing their nonlinear predictor performance by including a final stage of adaptive linear prediction, with the latter work showing gains of between 1.6 dB and 2.0 dB over the final linear stage by itself. We follow this point of view to improving our nonlinear predictor performance by linearly combining the three predictor outputs, resulting in the *nonlinear–linear* processing scheme described in the following text.

Linearly Combining Predictor Outputs for Improved Performance During the course of the experiment, we noted that the error sequences produced by an ensemble of nonlinear predictor outputs trained on a given speech sample with different parameters exhibit some residual correlation with the desired prediction. This observation suggested that, by standard properties of least-squares estimators, some further improvement in prediction performance should be possible when the predictor outputs are used as inputs in an additional level of regression on the desired (actual) speech signal. In selecting a compatible structure for this subsequent processing, it was desirable to retain as much as possible the recursive on-line nature of the algorithm without significantly increasing the computational burden. Thus the sixth line of the overall result tables shows the best observed performance for each speech sample when the three RBFN predictor outputs $\tilde{Y}_1(i)$, $\tilde{Y}_2(i)$, and $\tilde{Y}_3(i)$ at each time step i are formed into 3-tuples $\tilde{Y}(i) = [\tilde{Y}_1(i),\ \tilde{Y}_2(i),\ \tilde{Y}_3(i)]^{\mathrm{T}}$ and taken as regressive vector inputs into another exponentially weighted RLS predictor or *linear combiner* (to avoid confusion with the reference adaptive linear predictor). As before, the regressive orders and weights of the best such RLS linear combiners are given in parentheses below their performance figures and are chosen from trials conducted over the parameter ranges specified in Table 3.16. In most cases, only the most recent RBFN predictor outputs are necessary to provide a further nontrivial performance

Table 3.11 Overall Experimental Results for Speech Prediction, Samples f150 to f154

Network no.	f150	f151	f152	f153	f154
1	15.29	15.35	14.72	13.76	15.49
2	15.58	15.67	14.91	12.90	15.07
3	14.46	15.55	14.55	13.78	15.15
NL avg.	15.11	15.52	15.36	13.48	15.24
RLS	11.05	13.10	12.29	9.869	13.68
(auto)	(50, 0.999)	(44, 0.99)	(44, 0.99)	(50, 0.999)	(46, 0.99)
RLS	16.74	17.37	16.42	15.37	16.93
(NL)	(3, 0.999)	(1, 0.99)	(1, 0.98)	(3, 0.999)	(1, 0.99)
RLS	16.93	17.43	16.47	15.50	17.16
(NL + auto)	(4 + 6, 0.999)	(1 + 4, 0.99)	(2 + 6, 0.999)	(4 + 8, 0.999)	(2 + 48, 0.999)

Note: All PSNR in dB.

Table 3.12 Overall Experimental Results for Speech Prediction Example, Samples f155 to f159

Network no.	f155	f156	f157	f158	f159
1	18.02	15.60	15.42	11.74	15.38
2	18.14	15.97	15.95	12.40	14.39
3	17.78	15.89	16.19	13.06	16.15
NL avg.	17.98	15.82	15.85	12.40	15.31
RLS	16.08	11.85	13.15	10.09	14.26
(auto)	(42, 0.99)	(50, 0.999)	(50, 0.999)	(50, 0.999)	(50, 0.999)
RLS	19.86	17.45	17.51	13.96	17.48
(NL)	(1, 0.99)	(3, 0.999)	(1, 0.99)	(1, 0.999)	(1, 0.999)
RLS	20.60	17.70	17.84	14.29	17.83
(NL+auto)	(1 + 4, 0.99)	(4 + 8, 0.999)	(4 + 32, 0.999)	(3 + 8, 0.999)	(3 + 8, 0.999)

Note: All PSNR in dB.

Table 3.13 Summary of Gains in Experimental Results for Speech Prediction, Female Speech Samples

Gain	Min.	Avg.	Max.
NL avg. over RLS(auto)	1.05	2.67	4.06
RLS(NL) over NL avg.	1.06	1.70	2.17
RLS(NL + auto) over RLS(NL)	0.05	0.27	0.74
Total over RLS(auto)	3.48	4.63	5.88

Note: All figures in dB.

Table 3.14 Summary of Gains in Experimental Results for Speech Prediction, Male and Female Speech Samples

Gain	Min.	Avg.	Max.
NL avg. over RLS(auto)	−0.48	2.16	4.06
RLS(NL) over NL avg.	0.67	1.64	2.17
RLS(NL + auto) over RLS(NL)	0.05	0.39	0.78
Total over RLS(auto)	2.29	4.18	5.88

Note: All figures in dB.

gain averaging 1.64 dB over both the male and female speech samples. Augmenting the RBFN predictor output 3-tuples with autoregressive inputs drawn directly from the speech samples gives an additional small improvement of 0.51 dB for the male speech samples and 0.27 dB for the female speech samples, on average, for the best observed linear combiners. The exact performance figures for this nonlinear-linear input configuration are given in the seventh line of the tables, where the notation

Table 3.15 Trial Parameters for Reference Adaptive Linear Predictor[a]

RLS Parameter	Trial Range/Setting
$\boldsymbol{P}(0)$	$100\boldsymbol{I}$
$1-\rho$	$0, 10^{-6}, 10^{-4}, 10^{-3}, 0.01{:}0.01{:}0.20$
p	$2:2:50$

[a] $a : h : b$ denotes sequence from a to b inclusive sampled at h, $\boldsymbol{P}(0)$ is initial inverse of input correlation matrix, ρ is exponential weight.

Table 3.16 Trial Parameters for RLS Linear Combiner on RBFN Outputs Only[a]

RLS parameter	Trial range/setting
$\boldsymbol{P}(0)$	$100\boldsymbol{I}$
$1-\rho$	0, 10^{-6}, 10^{-4}, 10^{-3}, 0.01, 0.02
p	1 : 1 : 6

[a] $a : h : b$ denotes sequence from a to b inclusive sampled at h, $\boldsymbol{P}(0)$ is initial inverse of input correlation matrix, ρ is exponential weight.

in parentheses is (*nonlinear* 3-*tuple order* + *linear autoregressive order*, *RLS weight*). Table 3.17 lists the trial parameter ranges in this final configuration, for which the average performance gain over the RLS predictor for both male and female speech samples is 4.18 dB. We note that this average performance gain is approximately 2 dB greater than that reported in the relevant rows of Tables II to IV of (Baltersee and Chambers, 1998), although that study was limited to three speech signals. This gain naturally comes at the price of increased computational complexity, namely $\mathcal{O}(n^3)$ per time step, where n is the number of basis function, versus $\mathcal{O}(p^2)$ for the linear RLS predictor, where p is the linear autoregressive order. Whether the increased computational complexity of the regularized RBFN predictor over a linear one such as the RLS predictor is acceptable depends upon the intended application, but we should note that further gains in the nonlinear predictor's performance over the linear one should (at least in principle) still be possible since not all network parameters were fully optimized, for example, the bandwidth parameters.

Table 3.17 Trial Parameters for RLS Linear Combiner on Both RBFN Outputs and Autoregressive Inputs[a]

RLS parameter	Trial range/setting
$\boldsymbol{P}(0)$	$100\boldsymbol{I}$
$1-\rho$	0, 10^{-6}, 10^{-4}, 10^{-3}, 0.01, 0.02
p_{NL}	1 : 1 : 6
p_{auto}	2 : 2 : 50

[a] $a : h : b$ denotes sequence from a to b inclusive sampled at h, $\boldsymbol{P}(0)$ is initial inverse of input correlation matrix, ρ is exponential weight.

3.8 SUMMARY AND DISCUSSION

We have presented two theorems relating the normalized KRE to the regularized strict interpolation RBFN that justify its application to nonlinear time-series prediction. In the case of certain NLAR processes, we showed that minimizing the risk over the training set is asymptotically optimal in global MS prediction error, thereby demonstrating the key role regularization plays in the strict interpolation RBFN. We then introduced the idea of dynamically updating the strict interpolation RBFN parameters to account for the nonstationarity induced by time-varying regression functions via recursive algorithms that show interesting parallels to the standard RLS algorithm. As a practical demonstration of these ideas, we designed dynamically (per-step) updated regularized strict interpolation RBFNs for the nonlinear autoregressive one-step-ahead prediction of speech signals. Simulation experiments on a suite of several phonetically balanced male and female speech samples are conducted that show that the dynamic regularized strict interpolation RBFN predicts on average 2.2 dB better (as measured by the predicted signal-to-noise ratio) than a corresponding adaptive linear predictor trained with the exponentially weighted RLS algorithm. We also described how a simple linear combination of an ensemble of nonlinear predictor outputs (and the usual autoregressive inputs) via the RLS algorithm can yield a further average 2 dB improvement in prediction performance. These results demonstrate that this *composite linear–nonlinear* structure based on regularized strict interpolation RBFNs can effectively predict nonlinear processes with little added computational complexity while alleviating the difficulty of optimal model parameter estimation.

4

NONLINEAR STATE ESTIMATION

4.1 INTRODUCTION

The *Kalman filter* (Kalman, 1960) can be rightfully considered a landmark development in the theory and practice of optimum, that is, minimum mean-square error (MMSE), filtering. At the heart of the Kalman filter lies the notion of a *state-space model* of a *dynamical system* that allows an elegant *recursive* solution to the optimum *linear* filtering problem. For the general *nonlinear* filtering problem, such a recursive solution is not feasible, that is, practically computable, as we shall discuss further. In this chapter, we shall see how artificial neural networks (ANNs) in general and the regularized strict interpolation RBFN in particular can provide a practical data-based method for the problem of estimating the state of a nonlinear dynamical system. We begin with a brief review of the nonlinear state estimation problem.

4.2 PROBLEM DESCRIPTION

The classical *nonlinear state estimation* or *filtering* problem can be posited as follows: Suppose that we are given a dynamical system described by the continuous-time *stochastic differential equations (SDEs)*:

$$dX(t) = f(X(t))dt + dV(t) \qquad \text{(state)} \tag{4.1}$$

$$dY(t) = g(X(t))dt + dW(t) \qquad \text{(observation)} \tag{4.2}$$

where $\{X(t) \in \mathbb{R}^d : t \in \mathcal{T}\}$ and $\{Y(t) \in \mathbb{R}^p : t \in \mathcal{T}\}$ are the *state* and *observation/measurement* processes defined over some index set $\mathcal{T} \subseteq \mathbb{R}^+$ and driven by (possibly correlated) noise processes $\{V(t) \in \mathbb{R}^d : t \in \mathcal{T}\}$ and $\{W(t) \in \mathbb{R}^d : t \in \mathcal{T}\}$, respectively. For example, one common case is $\mathcal{T} = (0, \infty)$ and the initial state $X(0)$ has a given distribution. We would like to estimate with MMSE the current state $X(t)$ given the current and past observations $\{Y(s) : s \leq t,\ s, t \in \mathcal{T}\}$, that is, determine the conditional expectation $\tilde{X}(t) \triangleq \mathbb{E}[X(t) | \mathcal{F}^Y(t)]$, where $\mathcal{F}^Y(t)$ is the *filtration* of the process $\{Y(t)\}$. When $f : \mathbb{R}^d \mapsto \mathbb{R}^d$ and $g : \mathbb{R}^d \mapsto \mathbb{R}^p$ are known linear maps and $\{V(t)\}$ and $\{W(t)\}$ are *Wiener* processes so that their formal derivatives are wide sense stationary (WSS) *white noises* with known correlation matrices, the celebrated Kalman filter provides a recursive solution to the problem. When f and g are not linear (but still known), however, more general mathematical methods are required, which in practical terms calls for the *numerical solution* of the continuous-time SDE system (4.1) and (4.2). To set the stage for comparison with the ANN approach to the same problem, we mention the following salient points[1]:

1. The motivation behind the SDE approaches is to deduce conditions under which a recursive, finite-dimensional scheme can be used to compute the conditional probability distribution of the state $X(t)$ given the *filtration*[2] $\mathcal{F}^Y(t)$ produced by the observation process $\{Y(s) : s \leq t,\ s, t \in \mathcal{T}\}$, where the expectation is with respect to the $(d + p)$ dimensional probability space P corresponding to the Wiener processes $(V(t), W(t))$. A (MMSE) optimal point estimate of the state can then be obtained as the mean of this conditional distribution.
2. In general, the solution of Eqs. (4.1) and (4.2) results in conditions on *all* the higher-order conditional moments [of an appropriately transformed version of the state $X(t)$] and thus leads to an *infinite-dimensional* system, which is clearly undesirable in applications. Of course, studies also show that, under suitable restrictions, these

[1]For further details, the reader may refer to (Haykin et al., 1997c).
[2]See note 1 in Appendix A.4.

infinite-dimensional systems can be reduced to finite-dimensional systems.

3. A major practical difficulty with even the finite-dimensional schemes is that they are often *unimplementable*, in the sense that the algorithms cannot be used to generate numerical results with only minor modifications such as the truncation of an infinite domain and the introduction of an iterative method.[3]
4. *Pathwise*, that is, a.s., convergence is possible as well as *strong* or *weak* L_2, that is, MS, convergence with the SDE solution methods.[4] This contrasts with, for example, ANN function estimation problems where, for computational tractability, least-squares techniques lead to corresponding MS convergence.

The key point is that while the theoretical aspects of the nonlinear state estimation problem by the direct solution of the corresponding continuous-time SDEs have generated many algorithms, their application in practical problems has been limited by both their requirement for strong *a priori* knowledge of the underlying maps and noise characteristics and their difficulties in computationally feasible implementation [a complementary overview of factors inhibiting the widespread use of the SDE approach can be found in the Introduction of Lo (1995)].

4.3 ANN APPROACH

As with the continuous-time SDE approach, the ANN approach is concerned with estimating the conditional mean of the state given present and past observations. On the other hand, the different nature of ANNs results in some modifications to the SDE framework approach previously described:

1. Since ANNs have finite-dimensional domains, we must *time-discretize* the SDEs in Eqs. (4.1) and (4.2) governing the state transition and observations. We assume that after appropriately discretizing time, we have the state equation

$$X(i+1) = f(X(i)) + V(i) \qquad i \in \mathbb{Z}^+ (\text{state}) \qquad (4.3)$$

[3]See, for example, the discussion on p. 220 of Sun and Glowinski (1993).
[4]See note 2 in Appendix A.4.

where $X(i) \in \mathbb{R}^d$ is the state vector at time step i, $f(\bullet)$ is a vector-valued nonlinear function of its argument, and $V(i)$ is the process noise vector. Similarly, for the observation equation, we have

$$Y(i) = h(X(i)) + W(i) \qquad i \in \mathbb{Z}^+ \text{(observation)} \tag{4.4}$$

where $Y(i) \in R^p$, $h : \mathbb{R}^d \to \mathbb{R}^p$, and $V(i)$ is the observation noise vector.

2. As a result, the conditional expectation being estimated is that of the state with respect to a *finite* number m of present and past observations. This approximation is exact only if the conditional density is *Markov* to the truncation order m, otherwise the present and all past observations are, in theory, required. There is often, however, a practical limit on the performance improvements possible with increasing m, so this finite-memory assumption is not unreasonably restrictive.
3. Given the above, after suitable least-squares training, the ANN output function then estimates the *regression function* of the state with respect to the chosen m observations and, by plugging in the m most current observations $y_m(i) \triangleq [y(i), y(i-1), \ldots, y(i-m+1)]^T$, yields a *point estimate* of the actual state $x(i)$. All this stands in contrast to the continuous-time SDE case, where the output is an estimate of the conditional state density with respect to the available observations, that is, a *function*. Note that the ANN approach implicitly assumes that the regression function being estimated is *stationary* (or at least slowly varying) in the time index i, whereas the SDE approach usually assumes the same for the state transition and observation function.
4. By their nature as flexibly parameterized classes of functions, ANNs typically require a random sample or *training set* of the process paths, denoted by

$$T_{N,S} = \left\{(X(i,\omega), Y(i,\omega)) \in \mathbb{R}^d \times \mathbb{R} : \omega \in S, i = 1, 2, \ldots, N\right\} \tag{4.5}$$

where S is a random sample of length $\#S$ from the sample space Ω, in place of the SDE approach's *a priori* knowledge, for example, of the state transition and observation functions, their associated noise statistics, and the initial state distribution. Although the assumed availability of state sample paths may appear as problematic as the SDE approach's assumed knowledge of the underlying system functions, one can conceive of a scenario in which the system of interest is under control during the training phase and one is interested in estimating its state when such control is not possible. Not surprisingly, there is a price to pay for the generality of the ANN approach; for example, it is clear that, in general, because a given state transition or observation process sample path explores only a portion of its respective function domain, any finite training set, no matter how large, is not nearly as informative as knowing the actual underlying function. For ANN approaches to be both practical and successful, they must address this and the related design issues of appropriate network size and training algorithm complexity needed to construct a reasonable estimate.

4.4 REVIEW OF CURRENT APPROACHES

In this section we review several current ANN approaches to the state estimation problem. Among the early applications of ANNs to optimum nonlinear filtering is the work of Lo (1994, 1995). Here it is assumed that the state and observation equations are unknown, but that one could obtain enough sample data (in his notation) $\{(\mathbf{x}(t,\omega), \mathbf{y}(t,\omega)),\ t = 1,\ldots, T, \omega \in S\}$ to adequately capture the underlying statistics of the state and observation variables. Assuming stationarity, the following result is proven for two distinct *recurrent multilayer perceptron (RMLP)* (Perlmutter, 1989) architectures which are described below.

The first RMLP architecture of Lo (1995) is called the *neural filter with fully interconnected neurons (NFFN)* and, as its name suggests, has a fully recurrent input–output structure. Let t be the discrete-time variable, and let the weight from the ith input to the jth neuron in the first layer be ω^1_{ji}; let ω^2_{ji} be the weight of the output of the ith neuron into the jth hidden neuron, and ω^r_{ji} be the weight of the lagged feedback (by one time unit) from the ith neuron to the jth neuron (in the first input layer). Thus, the activation level $\beta_j(t)$ and the weighted sum $\eta_j(t)$ of the jth neuron satisfy

$$\beta_j(t) = g(\eta_j(t))$$

$$\eta_j(t) = \omega_{j0} + \sum_{i=1}^{m} \omega^1_{ji} y_i(t) + \sum_{i=1}^{q} \omega^r_{ji} \beta_i(t-1)$$

where g is a monotone increasing function such as tanh. The ith output $\alpha_i(t)$ is then given by

$$\alpha_i(t) = \sum_{j=1}^{q} \omega_{ji}^2 \beta_j(t) \tag{4.6}$$

for $i = 1, 2, \ldots, k$. The second RMLP architecture, called the *neural filter with ring interconnected neurons (NFRN)*, has a partially recurrent input–output structure as sketched in Fig. 4.1. Of the $i+j$ output nodes, $\alpha_1(t+1), \ldots, \alpha_k(t+1)$ are teacher forced, and $\beta_1(t+1), \ldots, \beta_j(t+1)$ are free outputs trained to have the teacher forced outputs. All the free outputs and the ith teacher forced outputs $\alpha_1(t+1), \ldots, \alpha_i(t+1)$ are delayed by one time unit before being fed into the input nodes $\beta_1(t), \ldots, \beta_j(t)$, $\alpha_1(t), \ldots, \alpha_i(t)$. In addition, the network has m input nodes on which the external inputs $\gamma_1(t+1), \ldots, \gamma_m(t+1)$ are clamped. Both of these networks are separately analyzed and the following result is deduced for both networks.

Theorem. *Consider the (discrete) random d-dimensional state process and p-dimensional observation processes* $\mathbf{x}(t)$ *and* $\mathbf{y}(t)$ *for* $t = 0, 1, \ldots, T$ *defined on a probability space* $(\Omega, \mathcal{F}, P)$. *Suppose that the range of* $\{\mathbf{Y}(t,\omega)|\omega \in \Omega\}_{t=1}^{T} \subset \mathbb{R}^p$ *is a.s. compact, and that* $\mathbb{E}[\mathbf{X}(t)^2] \leq \infty$ *for* $t = 0, \ldots, T$. *Then, if* $\boldsymbol{\alpha}(t)$ *is the network's output at time t, and* $\epsilon > 0$

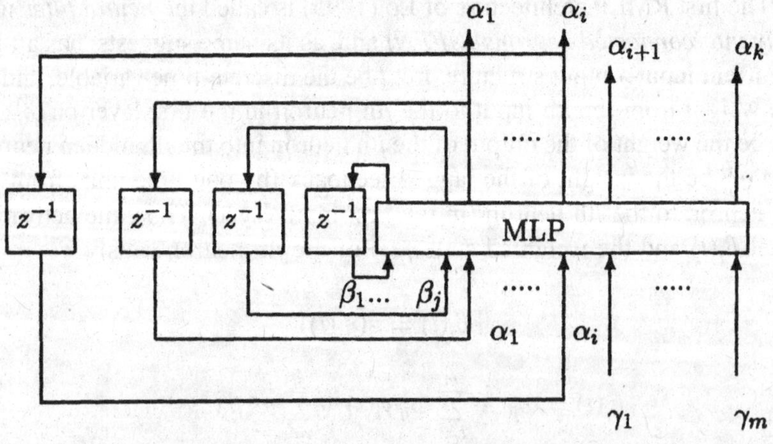

Figure 4.1 Input–output structure of NFRN (after Lo (1995)).

is given, there exists a sufficiently large RMLP such that

$$\frac{1}{T}\sum_{t=1}^{T} \mathbb{E}[\|\boldsymbol{\alpha}(t) - \mathbb{E}[\boldsymbol{X}(t)|\boldsymbol{Y}^T]\|^2] < \epsilon$$

where $\boldsymbol{Y}^{\mathsf{T}} \triangleq [Y(1), \ldots, Y(T)]$.

The theorem states that the RMLP architectures in question are sufficiently flexible to approximate the behavior of the desired conditional mean function in mean square to an arbitrary degree of accuracy over any given finite set of time points. While this theoretical result is necessary if the RMLP approach to optimum nonlinear filtering is to be a fruitful one, there are at least two distinct practical difficulties with implementing the conclusion of such an *existence* theorem, as we have elaborated in the Introduction:

1. For any nontrivial training set sizes and functions to be learned, the cost functions, which must be minimized over all possible network parameters during learning, are usually nonconvex and admit many *local minima*. Even if this optimization problem is alleviated, there remains the issue of the error induced in the "optimal" network parameters $\boldsymbol{w}^*$ when, as is common in many ANN learning schemes, these parameters are obtained by minimizing the *sample-based estimate*

$$\boldsymbol{w}^* \triangleq \inf_{\boldsymbol{w}} C(\boldsymbol{w}) \qquad C(\boldsymbol{w}) \triangleq \frac{1}{T(\#S)} \sum_{\omega \in S} \sum_{t=1}^{T} \|\boldsymbol{\alpha}(t, \omega) - \mathbb{E}[\boldsymbol{X}(t)|\boldsymbol{Y}^{\mathsf{T}}(\omega)]\|^2 \tag{4.7}$$

   (where $\#S$ is the number of samples per time point and $\boldsymbol{\alpha}(t)$ is the network output at time t) in place of the desired target cost function itself. It should again be mentioned, however, that both of these issues have been addressed in other work on ANN learning (White, 1989, 1990).

2. Given a finite training set, no guidelines are provided for selecting an appropriately sized network, for example, of n neurons, to yield the best *out-of-sample* or *generalization* performance. It is well known that without such guidelines, the problems of *overfitting* (too large an n) and *underfitting* (too small an n) can lead to poor generalization performance; for example, see Geman et al. (1992).

Nonetheless, a clear advantage of such an ANN approach is that no *a priori* knowledge of the statistics of the state and observation processes is required, other than having sufficient sample data to properly train the network (via temporal backpropagation). On the other hand, the larger T (the period of operation of the filter) is, the larger the network needed and correspondingly the greater the training time. Furthermore, the iterative optimization methods used are not particularly well suited to the incremental learning desired in a nonstationary environment.

In view of some of the shortcomings of the previous approach, we should mention the earlier work of Parsini and Zoppoli (1994, 1996). The first work presents nonrecursive as well as recursive techniques, which reduce the filtering problem described in Eqs. (4.3) and (4.4) to one of nonlinear optimization. Specifically, their recursive scheme involves sequentially minimizing n (where n is proportional to the observation period) functionals of the form:

$$J_i = \sum_{p=1}^{i} \phi_p(\|y_p - h_p(\hat{x}_p)\|) + \sum_{p=1}^{i} \psi_{p-1}(\|\hat{x}_p - f(\hat{x}_{p-1})\|) \qquad (4.8)$$

for $i = 1, 2, \ldots, n - 1$ (the case $i = 0$ is treated differently) where $\hat{\boldsymbol{x}}_i^p$ is an estimate of $\boldsymbol{x}_i$, and ϕ_p and ψ_p are arbitrary smooth increasing functions with $\phi_p(0) = \psi_p(0) = 0$. As compared with (4.7), these functionals are formulated according to what the authors call the *linear-structure preserving principle*, which is designed to emulate the linear structure of the Kalman filter. The actual solution to each minimization problem relies on the following procedure performed upon a suitable multilayer perceptron (MLP) [the $\boldsymbol{x}_i$ are the state variables as in Eq. (4.3) and the $\boldsymbol{y}_i$ are the observation variables as in Eq. (4.4); a variable with a p-superscript are the *predicted* versions of that variable]:

$$\hat{\boldsymbol{x}}_i = \hat{\boldsymbol{x}}_i^p + \tilde{\gamma}(\boldsymbol{y}_i - \boldsymbol{y}_i^p, \boldsymbol{\omega}_i) \qquad i = 0, 1, \ldots, n-1$$
$$\boldsymbol{x}_0^p = \boldsymbol{\alpha}$$

where $\boldsymbol{\alpha}$ is an *a priori* estimate of $\boldsymbol{x}_0$, and $\tilde{\gamma}(e_i, \boldsymbol{\omega}_i)$ is a multilayer feedforward network with weights $\boldsymbol{\omega}_i$ and inputs $e_i = y_i - y_i^p$. The flowchart for the scheme of Parisini and Zoppoli (1994) is reproduced in Fig. 4.2 (recall that v and w are the state and observation noises, respectively). Thus, at step $m + 1$, a nonlinear optimization is performed on the set of weights $\boldsymbol{\omega}_{m+1}$ of the $(m + 1)$st network while freezing the m previously computed weight vectors $\{\boldsymbol{\omega}_i\}_{i=1}^{m}$. This recursive implementation has the price of being structurally suboptimal (although not very

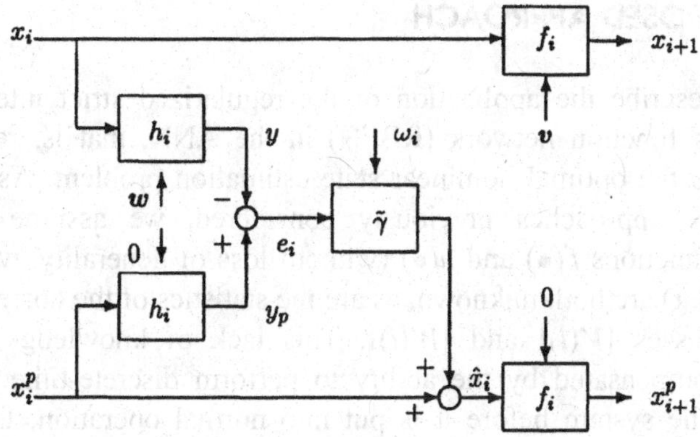

Figure 4.2 Flowchart for scheme of Parisini and Zoppoli (1994).

much so in practice, according to the authors) compared to the alternative nonrecursively implementable scheme laid out by the authors. Fortunately, in contrast to the approach of Lo (1995), the weights are adjusted by a gradient descent algorithm, since the assumed knowledge of the probability distribution functions of the state and observation noises allows the generation of "realizations" of a gradient function. Then by using standard backpropagation rules, these gradients may be computed for use in an appropriate weight update function until convergence is achieved. Other assumptions in their method are that the state and observation processes are zero mean, i.i.d., and mutually independent. As with (Lo, 1994, 1995), numerical simulations are performed on the problem of *bearings only measurement*, which is drawn from the more general class of *target motion analysis* problems. The test problem consists of an observer performing a series of tight maneuvers while acquiring noisy observations of the line of sight angle it makes with a target moving at constant speed. The results presented therein show significant performance gains over the extended Kalman filter, where it is known that the filter can diverge due to ill-conditioning of the covariance matrix.

Even with the *a priori* knowledge of the underlying statistics of the system, the recursive nature of the above scheme addresses neither the problem of excessive network complexity (when the observation period is large or has no *a priori* bound) nor the problem of actual network design (that is, how to determine the optimal network size and structure for a given observation period).

4.5 PROPOSED APPROACH

We now describe the application of the regularized strict interpolation radial basis function network (RBFN) in the ANN, that is, regression, approach to the optimal nonlinear state estimation problem. As with the other ANN approaches previously considered, we assume that the nonlinear functions $f(\bullet)$ and $h(\bullet)$ (without loss of generality, we assume h to be scalar) are both unknown, as are the statistics of the corresponding noise processes $\{V(i)\}$ and $\{W(i)\}$. This lack of knowledge is again partially compensated by the ability to perform discrete-time measurements on the system before it is put into normal operation. Unlike Lo (1994, 1995), however, under the stationary regression function assumption discussed in the previous section, we require only a single (sufficiently long) joint realization of the state and observation processes for network training, that is, S in Eq. (4.5) is a singleton, hence we simply write $T_N \triangleq \{(\mathbf{X}(i), Y(i)) \in \mathbb{R}^d \times \mathbb{R}\}_{i=1}^N$ where N is the sequence length and d is the dimension of the state vector $\mathbf{X}(i)$. Aside from this difference, the following issues from Chapter 3 bear repeating in the context of optimum nonlinear filtering:

1. For the regularized strict interpolation RBFN, Corollary 1 directly addresses item 1 of the discussion on the RMLP neural filter. The result also addresses item 2 by implying that the level of smoothing chosen by an appropriate cross-validation (CV) method asymptotically achieves a proper balance between the extremes of underfitting (excessive estimator bias) and overfitting (excessive estimator variance), since otherwise the MSE would not be minimized.
2. The training data are, in general, correlated from sample to sample, that is, *dependence* is present. For example, the discrete-state process $\{\mathbf{X}(i)\}$ constructed according to Eq. (4.3) is clearly dependent by the action of f (even when $\{V(i)\}$ is an i.i.d. process). From this it follows that the observation process $\{Y(i)\}$ in (4.4) is also dependent, for example, $Y(i)$ and $Y(i-1)$ are dependent. Because the MS consistency of neural estimators has been usually claimed only in the case of i.i.d. training data, the optimality of such estimators in the stochastic filtering context cannot be asserted without further study.
3. The processes $\{\mathbf{X}(i)\}$ and $\{Y(i)\}$ may be *nonstationary*. For general f in (4.3), the state process $\{\mathbf{X}(i)\}$ is clearly nonstationary [see, for example, p. 47 of Györfi et al. (1989)], hence the corresponding

observation process $\{Y(i)\}$ is also generally nonstationary. Once again, the i.i.d. data assumption underlying the majority of neural network studies in not applicable.

Thus, in theoretical terms, the regularized strict interpolation RBFN appears to be well suited to the regression approach to MMSE filtering or state estimation, in contrast with, for example, the RMLP. In practical terms, the regression estimate is implemented in the obvious way:

1. From T_N, derive the d component-wise training subsets of length $n = N - m + 1$ as $T_{n,k} \stackrel{\Delta}{=} \{(\boldsymbol{Y}_m(i), X_k(i)) \in \mathbb{R}^m \times \mathbb{R}\}_{i=m}^{N}$ for $k = 1, 2, \ldots, d$, where $X_k(i)$ is the kth component of the state vector $\boldsymbol{X}(i)$ and $\boldsymbol{Y}_m(i) \stackrel{\Delta}{=} [Y(i-j)]_{j=0}^{m-1}$.
2. For $k = 1, 2, \ldots, d$, obtain the kth regularized strict interpolation RBFN estimate $\tilde{f}_{n,k}$ as the solution to the interpolation problem specified by $T_{n,k}$ in the usual way. Note that with this construction, each network may have its own regularization parameter $\tilde{\lambda}_{n,k}$ (among others) chosen by CV on its particular training subset. Alternatively, one could simply aggregate the d interpolation problems of the form Eq. (15) in the Introduction together with a *single* regularization parameter λ_n as
$$(\boldsymbol{G}_n + \lambda_n \boldsymbol{I})\boldsymbol{\theta}_n = \boldsymbol{X}_n \tag{4.9}$$
where the kth column of $\boldsymbol{\theta}_n \in R^{n \times d}$ is the weight vector for $\tilde{f}_{n,k}$ and the kth column of $\boldsymbol{X}_n \in R^{n \times d}$ is $[X_k(i)]_{i=m}^{N}$. For the small d occurring in the subsequent experiments, we use the former method since in that case the additional computational burden is not overwhelming.
3. Form the overall network estimate $\tilde{\boldsymbol{X}}(i)$ for $\boldsymbol{X}(i)$, $i > N$, as
$$\tilde{\boldsymbol{X}}(i) = [\tilde{f}_{n,k}(\boldsymbol{Y}_m(i))]_{k=1}^{d} \tag{4.10}$$

4.6 EXPERIMENTAL RESULTS

In this section, we present two sets of experimental results. The results presented in Section 4.7 pertain to a comparison of the regularized strict interpolation RBFN approach to the SDE approach described in Sun and Glowinski (1993). The results presented in Section 4.8 pertain to a comparison of the regularized strict interpolation RBFN approach to the RMLP approach described in Lo (1994, 1995).

4.7 COMPARISON TO THE SDE APPROACH

Here we repeat Example 3 of Sun and Glowinski (1993) under modified conditions. The original SDE defining the system is

$$
\begin{aligned}
d\boldsymbol{X}(t) &= \begin{bmatrix} 2(\cos 2t - X_1^2(t) - X_2^2(t))\cos t \\ 2X_1(t) + (4 - X_1^2(t) - X_2^2(t))\sin t \end{bmatrix} dt + \epsilon_X\, d\boldsymbol{V}(t) \\
d\boldsymbol{Y}(t) &= \begin{bmatrix} 2\sin(X_1(t)) \\ 2X_2^2(t) \end{bmatrix} dt + \epsilon_Y\, d\boldsymbol{W}(t)
\end{aligned} \tag{4.11}
$$

where $\boldsymbol{X}(t) \triangleq [X_1(t), X_2(t)]^{\mathsf{T}}$ and with parameters $\epsilon_X = \epsilon_Y = 0.1$, $\mathbb{E}[\boldsymbol{X}(0)] = [0, 0]^{\mathsf{T}}$, $\boldsymbol{Y}(0) = [0, 0]^{\mathsf{T}}$, and $t \in [0, \pi/2]$. The initial state $\boldsymbol{X}(0)$ is distributed as the standard normal about $\mathbb{E}[\boldsymbol{X}(0)]$. We discretize the SDE in time using a simple forward Euler scheme with $\Delta t = 2\pi/(n-1)$ and $n = 10N$, where N is number of training pairs to be obtained by subsampling the full state and observation sample paths at rate 10. The discretized vector noise differentials $d\boldsymbol{V}(t)$ and $d\boldsymbol{W}(t)$ are sequentially simulated by using componentwise independent samples from a normal pseudo-random number generator with zero mean and variance Δt. To avoid numerical instability with the crude discretization scheme used, we set $\epsilon_X = \epsilon_Y = 0.05$ and fixed the initial state at $\mathbb{E}[\boldsymbol{X}(0)]$; as will be seen, even these choices result in discretized state sample paths with somewhat greater spatial variability than that shown in Figure 3 of Sun and Glowinski (1993).

We follow the strict interpolation RBFN design procedure detailed in the previous section. The only change is that because we have two-dimensional *vector* observations, the *effective* input vector is formed by conçatenating m observation vectors into a single $2m$-long vector as $\boldsymbol{y}_{2m}(i) \triangleq [\boldsymbol{y}^{\mathsf{T}}(i), \boldsymbol{y}^{\mathsf{T}}(i-1), \ldots, \boldsymbol{y}^{\mathsf{T}}(i-m+1)]^{\mathsf{T}}$. The result is two scalar output networks $\tilde{f}_{n,1}$ and $\tilde{f}_{n,2}$, which estimate the regression of the discretized state components $X_1(i)$ and $X_2(i)$ with respect to a common input vector $\boldsymbol{Y}_{2m}(i)$. The vector regression order m is determined for $\tilde{f}_{n,k}$, $k = 1, 2$, by setting their RBFN input norm-weighting matrices to $\boldsymbol{U}_{n,k}^{-1} \triangleq \mathrm{diag}[\alpha_{k,j}\sigma_{k,j}^2]_{j=1}^m$, where $\sigma_{k,j}^2$ is the sample variance of the jth input variable over $T_{n,k}$. These *input scaling parameters* $\alpha_{k,j}$ are estimated from the training data $T_{n,k}$ along with the regularization parameter $\lambda_{n,k}$ by the GCV procedure. The approximate minimization of the GCV criterion function is performed by the MATLAB version 4.2c Optimization Toolbox routine `constr`, which implements a simplex-type method of nonlinear multivariate minimization over a quadrantal region specified

Table 4.1 Optimization Settings for constr Routine in Approximate GCV Minimization[a]

Optimizer Settings	
OPTIONS(2) (input termination tolerance)	10^{-6}
OPTIONS(3) (function termination tolerance)	10^{-6}
OPTIONS(14) (max. no. of iterations)	1000

Variable Search Settings

Input Variable	Min.	Init.	Max.
λ	10^{-8}	0.001	1
α	10^{-4}	0.5	10^{8}

[a]Settings for α apply to each input scaling parameter.

by upper and lower bounds on the input variables. The relevant settings used in this problem are listed in Table 4.1 (optimizer settings not listed are kept at their default values).

Because of the inverse weighting, scalar observations (input components) corresponding to the larger estimated input scaling parameters have less influence on the network output (for a given input) than those associated with smaller input scaling parameters, so it appears reasonable to set m just large enough to include these values. What is most interesting is that even as m is increased, the smallest input scaling parameters for both networks are associated with only the most recent observation vector $y(i)$, with all other input scaling parameters at least an order of magnitude larger, that is, less significant. A typical example of this phenomenon for $m = 2$ (albeit for $\epsilon_x = \epsilon_y = 0.1$) can be found in Table 4.2.

Given that $m = 1$ appears sufficient, the final network input scaling and regularization parameters as determined by the GCV procedure are listed in Table 4.3.

The generalization performance of the two networks is tested against 20 other state sample paths generated independently in the same fashion as

Table 4.2 Example of GCV-Selected Input Scaling Parameters for $m = 2$ and $\epsilon_x = \epsilon_y = 0.1$

Estimate	$\alpha_{k,1}$	$\alpha_{k,2}$	$\alpha_{k,3}$	$\alpha_{k,4}$
$\tilde{x}_1(i)$ $(k = 1)$	0.2568	1.990	2.889	1.906
$\tilde{x}_2(i)$ $(k = 2)$	2.008	0.1205	1.240	1.039

Table 4.3 GCV-Selected Parameters for Final Network, SDE Comparison

Estimate	$\alpha_{k,1}$	$\alpha_{k,2}$	λ
$\tilde{x}_1(i)$ $(k = 1)$	0.1551	0.9091	0.004085
$\tilde{x}_2(i)$ $(k = 2)$	0.3480	2.578	0.001449

the training data; some representative results are presented in Figs. 4.3 to 4.6, in approximate order of increasing performance as measured by the root MSE (RMSE) for each state component. Although the reader is invited to compare these figures with the corresponding Figure 3 of Sun and Glowinski (1993), the substantially different nature of our approach, that is, regression, compared to that of Sun and Glowinski (1993), that is, pathwise-convergent numerical solution of SDEs, means due caution should be exercised in drawing any definitive conclusions, especially in view of the limited scope of the experiment. We should also mention that a comparison to an extended Kalman filter was attempted but was not successful due to difficulties in obtaining an appropriate equivalent discrete-time system for the continuous-time nonlinear system (4.11);

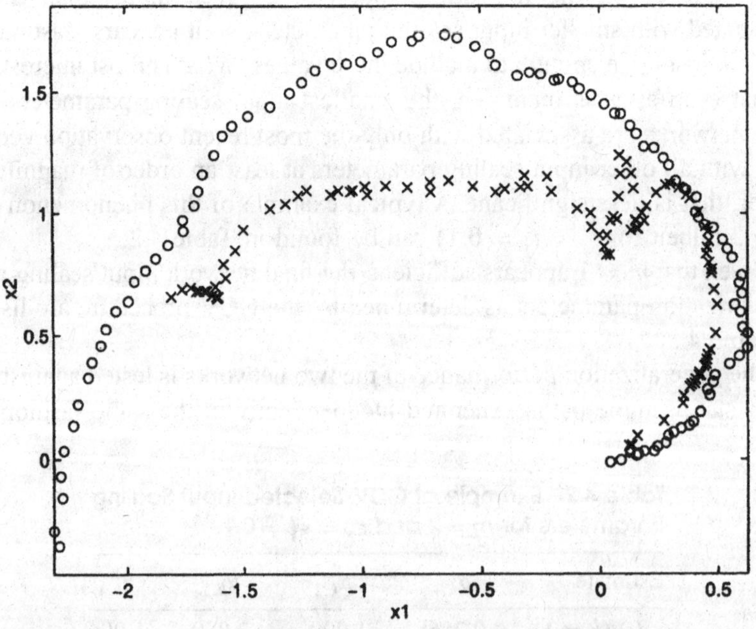

Figure 4.3 Example of estimated (x) versus actual (o) state sample path, RMSE $x_1 = 0.3423$, RMSE $x_2 = 0.3940$.

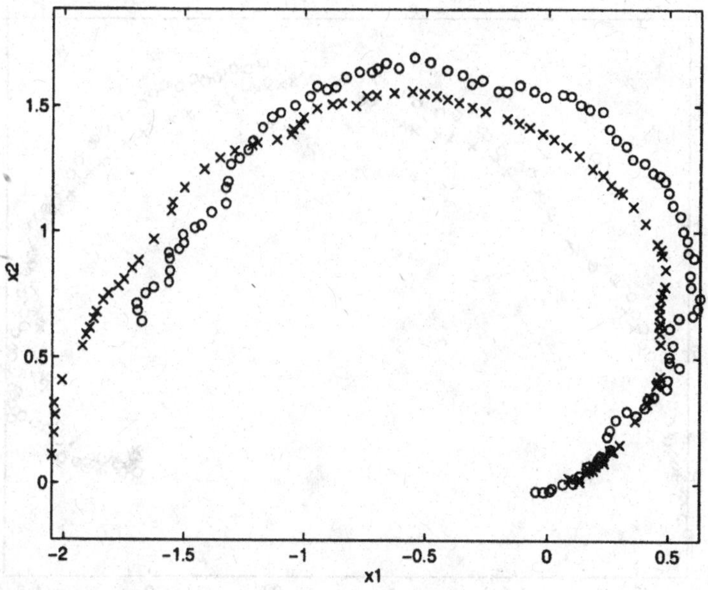

Figure 4.4 Example of estimated (x) versus actual (o) state sample path, RMSE $x_1 = 0.1762$, RMSE $x_2 = 0.1950$.

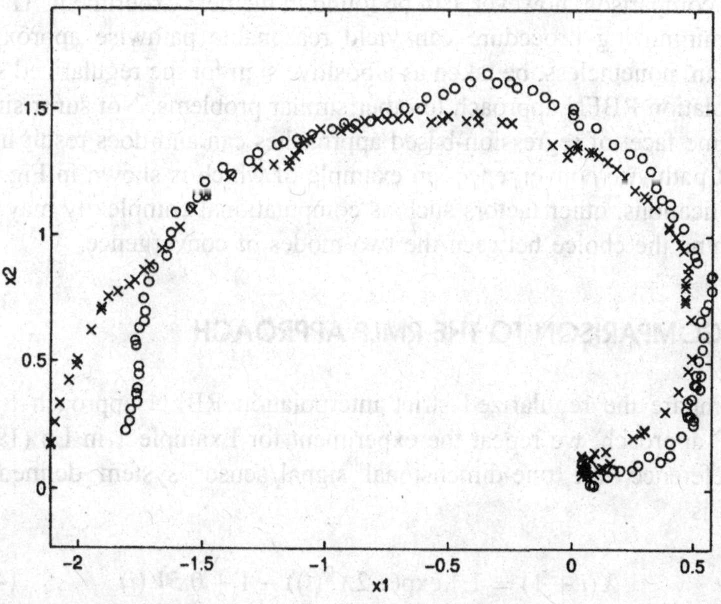

Figure 4.5 Estimated (x) versus actual (o) state sample path, RMSE $x_1 = 0.1757$, RMSE $x_2 = 0.0967$.

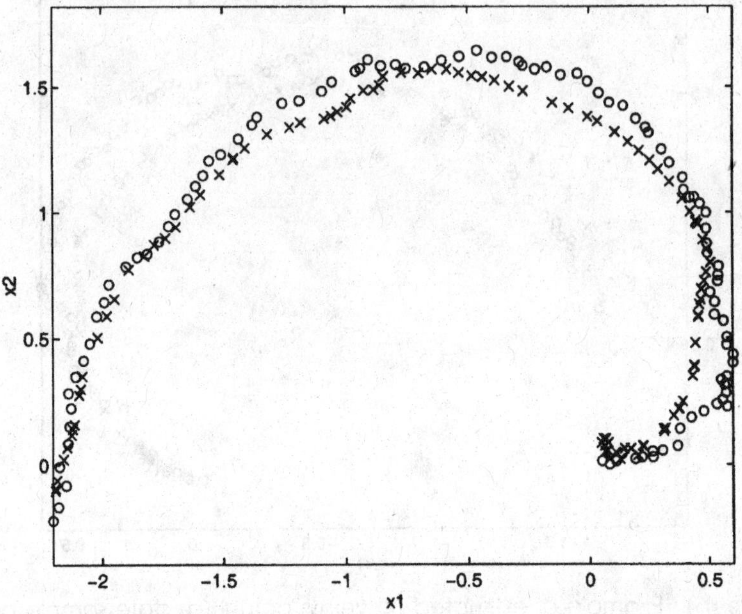

Figure 4.6 Example of estimated (x) versus actual (o) state sample path, RMSE $x_1 = 0.1252$, RMSE $x_2 = 0.1191$.

such a comparison, however, can be found in the next experiment. That an MSE-minimizing procedure can yield reasonable pathwise approximations can, nonetheless, be taken as a positive sign for the regularized strict interpolation RBFN approach to other similar problems. Not surprisingly, this same facet of regression-based approaches can and does result in the loss of pathwise convergence, an example of which is shown in Fig. 4.3. In applications, other factors such as computational complexity may well determine the choice between the two modes of convergence.

4.8 COMPARISON TO THE RMLP APPROACH

To compare the regularized strict interpolation RBFN approach to the RMLP approach, we repeat the experiment for Example 1 in Lo (1995). For reference, the one-dimensional signal/sensor system defined for $i \in \mathbb{Z}^+$ is

$$X(i+1) = 1.1 \exp(-2X^2(i)) - 1 + 0.5V(i) \tag{4.12}$$

$$Y(i) = X^3(i) + 0.1W(i) \tag{4.13}$$

where $X(0)$ is Gaussian with mean -0.5 and variance 0.1^2. The signal/sensor noises $\{V(i)\}$ and $\{W(i)\}$ are statistically independent, standard white Gaussian sequences with zero mean and unit variance.

Figures 4.7 and 4.8 show that even with as few as $N = 100$ training data, f and h can be estimated from T_N with fair accuracy. As before, each regularized RBFN network has a Gaussian kernel $K(r) = \exp(-r^2/2)$, $r \in \mathbb{R}^+$, with diagonal norm-weighting matrix of the form $U_n^{-1} \triangleq \mathrm{diag}[\alpha_j \sigma_j^2]_{j=1}^d$, where $d = 1$ is the state dimensionality and the σ_j^2 is the sample variance of the jth input variable over T_N. These relative input weights and the regularization parameter λ_n are again determined from T_N by approximately minimizing the GCV criterion using the same optimizer settings as in Table 4.1, with results shown in Table 4.4.

Naturally, the quality of the estimation improves in the region containing a higher density of training points than at the outliers in the training set; this aspect is most clear in Figure 4.7, where the outlier near $x(i-1) = -2.5$ has an exaggerated effect on the shape of the estimated curve.

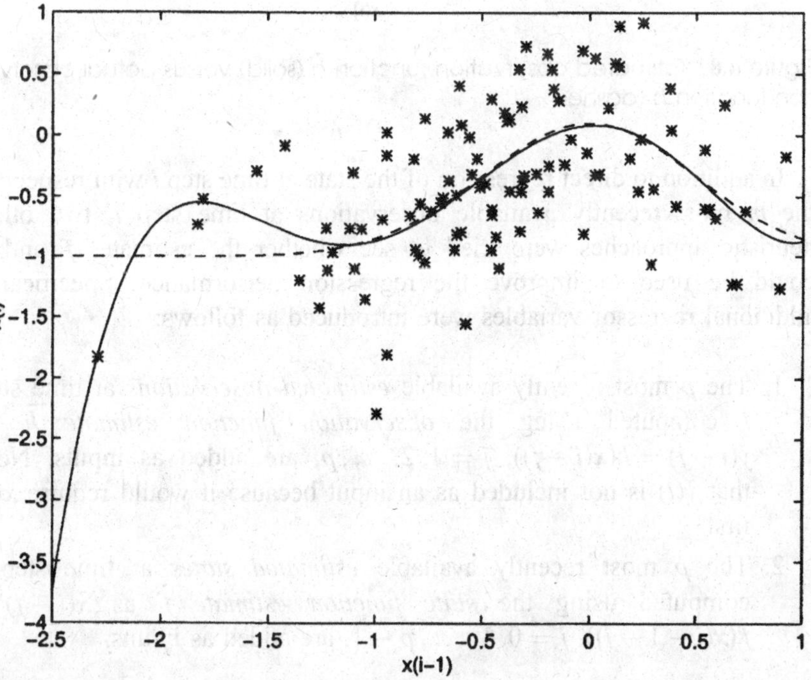

Figure 4.7 Estimated transition function $\tilde{f}$ (solid) versus actual transition function f (dashed).

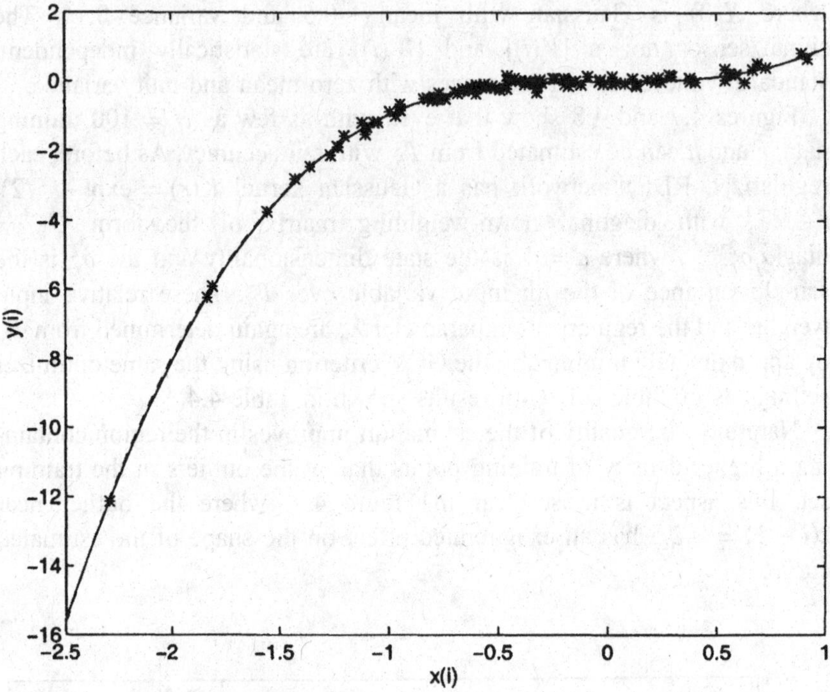

Figure 4.8 Estimated observation function $\tilde{h}$ (solid) versus actual observation function h (dashed).

In addition to direct regression of the state at time step i with respect to the m most recently available observations at time step i, two other heuristic approaches were tried to see whether the estimates $\tilde{f}$ and $\tilde{h}$ could be used to improve the regression performance. Specifically, additional regressor variables were introduced as follows:

1. The p most recently available *estimated observations* at time step i computed using the *observation function estimate* $\tilde{h}$ as $\tilde{y}(i-j) = \tilde{h}(\boldsymbol{x}(i-j))$, $j = 1, 2, \ldots, p$, are added as inputs. Note that $\tilde{y}(i)$ is not included as an input because it would require $\boldsymbol{x}(i)$ first.
2. The p most recently available *estimated states* at time step i computed using the *state function estimate* $\tilde{f}$ as $\tilde{\boldsymbol{x}}(i-j) = \tilde{f}(\boldsymbol{x}(i-1-j))$, $j = 0, 1, \ldots, p-1$, are added as inputs.

Our simulations indicated no statistically significant improvement in estimation performance with these possible additional inputs, even with increased m and p. In fact, the scaling factor α_1 selected by the GCV criterion for the input variable $y(i)$ is much smaller, that is, indicates greater significance, than the other α_j for both the other approaches with the additional regressor variables. For example, using $m = 2$ for $\tilde{\mu}$ gives $\alpha_1 = 0.5746$ for $y(i)$, while $y(i-1)$ is assigned $\alpha_2 = 274.5$. That this observation holds across the different m and p for the other approaches strongly suggests that most of the information about $x(i)$ is contained in $y(i)$, that is, $m = 1$ would be sufficient. As a representative result, Fig. 4.9 shows the average RMSE over 1000 test sequences of 120 time points with $N = 800$ training data and a regressive order of $m = 2$ for the observations. Visually, this figure appears quite similar to that in Figure 2.4 of Lo (1994, 1995) for the RMLP neural filters. A summary of the numerical results in comparison to the original results of Lo (1994, 1995) for the same problem can be found in Tables 4.4 and 4.5.

Although the mean RMSE of 0.2260 (with standard deviation of 0.0111) over the 120 time points is somewhat larger than the 0.2120 and 0.2122 reported for the two neural filters in Figure 2.4 of Lo (1995), one should keep in mind that the number of training data $N = 800$ we

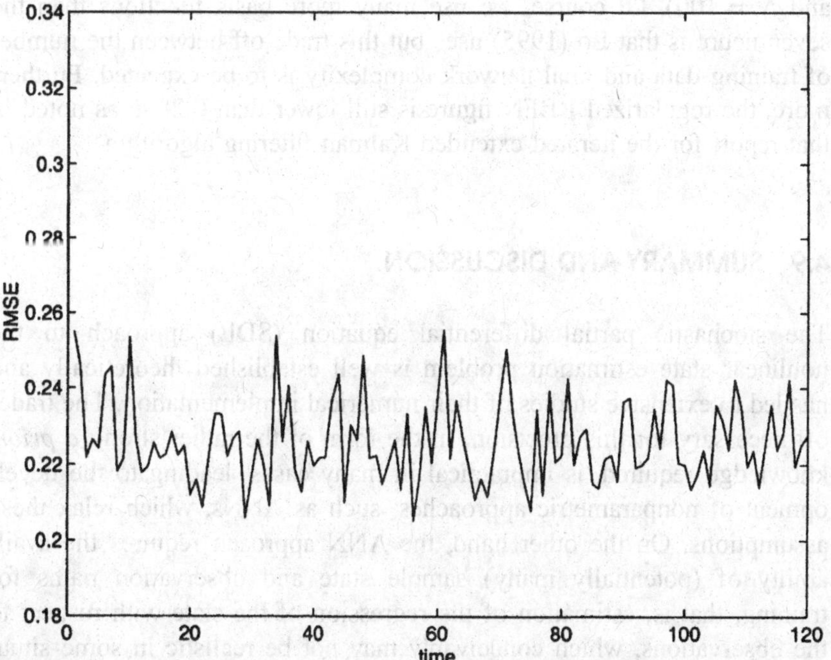

Figure 4.9 State estimate RMSE over 1000 test sequences of 120 time points.

Table 4.4 GCV-Selected Parameters for Final Network, RMLP Comparison

Estimate	α_1	α_2	λ
$\tilde{x}(i)$	0.5746	274.5	0.05228

Table 4.5 Comparison of State Estimate RMSE over 120 Time Points for System Defined by (4.12) and (4.13)

Network	RMSE of state estimate
RBFN	0.2260
IEKF (Lo)	0.2806
NFFN (Lo)	0.2120
NFRN (Lo)	0.2122

used is much smaller than the 200,000 training samples effectively used in the iterative algorithm of Lo (1995) 100 sweeps or *epochs* over $\#S = 200$ and $N = 100$). Of course, we use many more basis functions than the seven neurons that Lo (1995) uses but this trade-off between the number of training data and final network complexity is to be expected. Furthermore, the regularized RBFN figure is still lower than 0.2806 as noted in that report for the iterated extended Kalman filtering algorithm.

4.9 SUMMARY AND DISCUSSION

The stochastic partial differential equation (SDE) approach to the nonlinear state estimation problem is well established theoretically and has led to extensive studies of their numerical implementation. The trade-off necessary for this precision, in the form of the rather strong *a priori* knowledge required, is impractical in many cases, leading to the development of nonparametric approaches, such as ANNs, which relax these assumptions. On the other hand, the ANN approach requires the availability of (potentially many) sample state and observation paths for training, that is, estimation of the regression of the state with respect to the observations, which conceivably may not be realistic in some situations. Barring the computational complexity of the SDE methods, it is clear that when strong prior knowledge is available, one should exploit it fully by using the model-based SDE approach to generate parsimonious solutions with known properties. If such prior knowledge is not available but sufficiently many representative sample paths are, then the first experiment, the comparison to the SDE solution method of Sun and Glowinski (1993), suggests that ANN regression-based methods can offer a useful alternative, keeping in mind the key differences of function versus point estimate and pathwise versus MS convergence between the two approaches.

On the basis of the limited simulation results obtained, it would be clearly premature to claim that either the regularized strict interpolation RBFN (with asymptotically optimal regularization parameter sequence) or the RMLP-based method provides superior performance among the ANN-based regression approaches to optimum, that is, MMSE, nonlinear state estimation. Nonetheless, the relatively stronger theoretical support for the practical techniques used in regularized strict interpolation RBFN design is a significant advantage when compared to the uncertainties surrounding the theoretical effectiveness of the design methods used in the RMLP-based approach. We should also note that, from a computational standpoint, the comparatively simplified training of regularized strict interpolation RBFNs also admits the possibility of applying one of the reduced complexity recursive update algorithms detailed in Section 3.6, while the RMLP-based method does not easily do so. Such updating can be important when the state transition function f and the observation function h are time variant.

Despite the initial success of the ANN-based regression approaches to the nonlinear state estimation problem, many open questions remain to be answered as listed below:

1. What are the particular features of the regression induced by the specific structure in the coupled dynamics of Eqs. (4.3) and (4.4)? It seems intuitive that the induced regression should admit more structure than for an arbitrary regression problem. If such features do indeed exist, how can they then be exploited to obtain improved performance or otherwise aid in estimator design, for example, selecting an appropriate input order m?
2. Can the preliminary state transition estimate $\tilde{f}_n$ and observation function estimate $\tilde{h}_n$ be exploited in the regression approach to improve performance? In the comparison to the RMLP-based method, we did not observe any significant change in estimation performance when $\tilde{f}_n$ and $\tilde{h}_n$ were used in the "obvious" way to generate additional regressor variables in the form of estimated

states or observations. It would be useful to develop some theory to account for this behavior in more general situations.

3. Is it possible to implement the optimal state estimate in a structure that parallels the linear Kalman filter? For example, under what conditions would an MMSE optimal state estimate $\tilde{X}^*(i) \triangleq \mathbb{E}[X(i)|Y_m(i)]$ be expressible in the *predictor–corrector* form

$$\tilde{X}^*(i) = \tilde{f}_n(\tilde{X}^*(i-1)) + G(Y_m(i)) \tag{4.14}$$

and how could such a G (if it exists) be estimated from the training data in T_N? Although equivalent in performance (at least in principle) to the direct regression of the state with respect to the observations, such a construction would have the advantage of efficient recursive computability. Variations on (4.14) include allowing general F in place of $\tilde{f}_n$ and ascertaining conditions under which the argument $Y_m(i)$ could be replaced by the *innovations* or *a priori errors* $Y(j) - \tilde{Y}(j)$, $j = i, i-1, \ldots, i-m+1$, as in the linear Kalman filter.

4. When f or h is nonstationary, that is, time varying, which procedures should be used to ensure that the state estimates continue to adapt to or *track* the actual state (with MMSE)? In view of the discussion above, it would be particularly advantageous if these updates could be made recursive while maintaining the predictor-corrector structure (4.14).

In applications where our knowledge of the system dynamics is practically limited, nonparametric techniques such as those based on ANNs hold much promise. Of these ANN-based regression approaches to the nonlinear state estimation problem, the regularized strict interpolation RBFN stands out as being computationally feasible as well as theoretically supported in its application. As the issues raised above are clarified, it is expected that RBFN-based ANN approaches will have a significant role to play in providing efficacious solutions to the nonlinear state estimation problem.

5

DYNAMIC RECONSTRUCTION OF CHAOTIC PROCESSES

5.1 INTRODUCTION

Chaotic dynamical systems have emerged recently as an important theory for understanding the possible mechanisms behind observed time signals. Indeed, one may say that the theory attempts to describe seemingly *random*, complex behaviors as the product of a set of "simple" (in a well-defined sense), *deterministic* coupled differential equations. Due to the breadth of the field, we shall necessarily confine our attention to those aspects of chaos that are absolutely relevant in understanding the application of regularized strict interpolation radial basis function networks (RBFNs) to the *dynamic reconstruction* of chaotic systems. For further background, the reader may consult one of several possible general references (Abarbanel, 1996; Baker and Gollub, 1996; Ott, 1993; Ott et al., 1994). Certain recent works, among them Scargle (1992) and Tong (1992), argue that tools from statistical time series are still relevant in the chaotic context, a viewpoint that we shall also reflect in our approach. We begin by setting the basic theoretical framework for the dynamic reconstruction of chaotic systems.

5.2 PROBLEM DESCRIPTION

In a generic sense, the problem of the dynamic reconstruction of chaotic systems is fairly straightforward to state: given a finite length sample t_N of a real-valued time series $\{x(i)\}_{i=1}^{N}$, estimate a function $\tilde{f}: \mathbb{R}^p \mapsto \mathbb{R}$ for some given $p > 0$ such that the *recursively forward iterated (RFI)* predicted sequence $\{\tilde{x}(i)\}_{i=N+1}^{\infty}$ formed by feeding successive RFI estimates back into the function input as

$$\tilde{x}(i+1) \stackrel{\Delta}{=} \tilde{f}(\tilde{\boldsymbol{x}}_p(i)) \tag{5.1}$$

where

$$\begin{aligned} \tilde{\boldsymbol{x}}_p(i) &\stackrel{\Delta}{=} [\tilde{x}(i), \tilde{x}(i-1), \ldots, \tilde{x}(i-p+1)]^{\mathsf{T}}, \qquad i = N+1, N+2, \ldots \\ \tilde{x}(i) &\stackrel{\Delta}{=} x(i) \qquad i = N-p+1, N-p+2, \ldots, N \end{aligned} \tag{5.2}$$

(approximately) reproduces given chaotic characteristics of the system responsible for the generation of the sequence $\{x(i)\}_{i=1}^{N}$. Where matters become more involved are, of course, in the chaos theory behind the interrelated issues of the choice of p, the existence and properties of the function $\tilde{f}$, and the definition of the characteristics to be reproduced. These issues will be addressed as needed to clarify their role in the design of regularized strict interpolation RBFNs for dynamic reconstruction. Even without delving into these technical details, it is already clear that the objective here is different in character from that of the previous chapters:

1. In the previous chapters, the main concern was minimizing the (global) mean-square *error* (MSE) of the estimate $\tilde{f}$ in approximating a function f assumed to relate the training input and output data [perhaps implicitly, as in the case of stochastic T_n and minimum MSE (MMSE) estimation]. For dynamic reconstruction, the emphasis is on the *reproduction* of particular *features* of the chaotic dynamical system underlying the training data, for example, local and global *attractors* in the *phase space*. While the former is necessary for the latter, it is clearly not sufficient, in the sense that two estimates with the same level of MSE can have radically different phase portraits compared to the original chaotic system.
2. In the regression context, the unknown function f being estimated was well defined in the sense that given stationary processes $\{\boldsymbol{Z}(i)\}$

and $\{Y(i)\}$, the conditional expectation function $f(\bullet) \triangleq \mathbb{E}[Y(i)|\boldsymbol{Z}(i) = \bullet]$ is a.s. unique, that is, if g is another version of the conditional expectation, then $f = g$ a.s.–P_Z. On the other hand, it is intuitive that the function f used in dynamic reconstruction is not unique, that is, there may exist more than one function whose RFI predicted time series captures the key chaotic characteristics of the original chaotic system. In fact, the key theorems supporting the dynamic reconstruction approach, for example, *Takens' embedding theorem* (Takens, 1981), merely assert (under suitable conditions) the *existence* of a smooth map f whose RFI predicted time series has the required chaotic properties.

3. The RFI operational mode of the estimate $\tilde{f}$ is unusual for the regularized strict interpolation RBFN, given its design by simultaneous interpolation for the one-step-ahead prediction problem described in Chapter 3, which is a *nonsequential* optimization procedure. The obvious error measure to use in place of $\tilde{J}_2$ [as defined in Eq. (5) in the Introduction] is the *average squared RFI prediction error* over t_N, namely

$$\rho(\tilde{f}, t_N) \triangleq \frac{1}{n} \sum_{i=p+1}^{N} (x(i) - \tilde{f}(\tilde{\boldsymbol{x}}_p(i)))^2 \tag{5.3}$$

whose minimization (with respect to the network parameters in $\tilde{f}_n$) requires a *sequential* procedure due to the recursive construction of the input vector $\tilde{\boldsymbol{x}}_p(i)$ at each time step i. The intuition that such RFI prediction-based cost functions are more appropriate for dynamic modeling than simultaneous one-step-ahead prediction-based cost functions is borne out in several simulation studies, for example, Principe et al. (1992) and Principe and Kuo (1995), among others. While there is no theoretical barrier to the application of such procedures to RBFNs, several issues would have to be addressed:

(a) The development of analogous cross-validation (CV) theory and techniques for the choice of λ (and possibly other parameters). It should be noted that, as a heuristic, one can select λ on the basis of the RFI prediction error (in a suitable norm) over a holdout set that is distinct from the training data used to define the network. This heuristic parameter selection method, which we shall call *cross-validation via iterated prediction (CVIP)*, did not improve performance significantly for the Lorenz and sea clutter reconstruction experiments which are described further on.

(b) The computational complexity, which can be geometrically greater than that of solving the strict interpolation (SI) linear algebraic equations. For example, at least $\mathcal{O}(Mn)$ function evaluations are required for M iterations over the n training pairs in T_N for the former case versus $\mathcal{O}(n^3)$ arithmetic operations for the latter case. Because of the lack of an *a priori* bound on M, it can be much larger than n^2.

(c) The resultant (global) *stability* of the closed-loop system generated by operating in the RFI predictive mode. Such stability is not, in general, a natural by-product of either standard simultaneous (regularized) least-squares or the proposed design procedure based on sequential optimization in the RFI predictive mode. Note that while sufficient, accurate reproduction of the attractors (for example, fixed points and orbits in the corresponding system phase space) by the estimate $\tilde{f}_n$ is clearly not necessary to ensure such stability.

The elementary state of our knowledge regarding the precise effects of these factors should therefore be kept in mind when evaluating the performance of the regularized strict interpolation RBFN in the following experiments. That said, the experiments do show that the regularized strict interpolation RBFN can form the basis for an effective dynamic reconstruction technique, particularly when the training data are *noisy*. As we will elaborate further on, this critical aspect of any realistic dynamic reconstruction problem has not to date been satisfactorily addressed by existing methods, which typically rely on *ad hoc* prefiltering of the training data.

5.3 REVIEW OF CURRENT APPROACHES

Roughly speaking, methods used in the dynamic reconstruction of chaotic processes can be divided into two classes (Casdagli, 1989):

1. *Global Methods*: these construct a single approximating function $\tilde{f}$ for all training data in T_N. Classical examples include polynomials and their rational compositions.
2. *Local Methods*: these construct the approximation $\tilde{f}$ by concatenating functions with *localized* support. More precisely, the prediction

at some $\boldsymbol{x} \in \mathbb{R}^p$ in the domain of $\tilde{f}$ is computed by a function constructed only with data in the neighbourhood of $\boldsymbol{x}$. Some well-known examples of such methods include nearest-neighbor methods and their generalized kernel extensions along with piecewise polynomials (for example, cubic splines).

Each class of methods has its own particular advantages and disadvantages; for example, global methods offer the possibility of parsimonious (in the number of function parameters) representations but in practice, without significant prior knowledge, one must often resort to flexible nonparametric methods with a large number of free parameters whose solution is computationally intensive. On the other hand, local methods involve functions that may be simpler to estimate from the training data over each local region but again, without some prior knowledge, the number of localized functions required can be large, leading to a large run-time network. Strictly speaking, the regularized strict interpolation RBFN with Gaussian kernel, which we shall be using, is a global method since the support of each network basis function is all of $\mathbb{R}^p$, but it is often effectively considered a local method, since each basis function models primarily the region about its center (which is an input datum in T_N). Indeed, this hybrid nature is one of the factors that led Casdagli (1989) to an early application of nonregularized RBFNs to the dynamic reconstruction problem. That work shows that, compared with polynomial and multilayer perceptron (MLP)-based predictors, nonregularized RBFNs can predict well in short-term one-step-ahead (OSA) mode, that is, the prediction of $x(i+j)$ with $\tilde{x}(i+j) \overset{\Delta}{=} \tilde{f}(\boldsymbol{x}(i+j-1))$, $j = 1, 2, \ldots, M$ for small M, as well as the more difficult RFI mode. The work notes, however, degraded prediction performance and, more importantly, poor recovery of key dynamical invariants for the nonregularized RBFN when white Gaussian noise is added to the time series considered. While stating that this shortcoming can be addressed by a variety of techniques, the work does not specifically examine nor test any one of those techniques. Similarly, in the global modeling method of Principe et al. (1992) and later Principe and Kuo (1995) using MLPs whose parameters are estimated by least-squares minimization of the RFI prediction error, the potential deleterious influence of noise on the quality and accuracy of reconstruction is not considered. We shall later give an example that clearly demonstrates the importance of regularization to strict interpolation RBFN predictor performance in dynamic reconstruction from noisy observations of chaotic processes.

5.4 EXPERIMENT: RECONSTRUCTION OF THE LORENZ SYSTEM FROM NOISY DATA

5.4.1 Results for Noise-Free Case

The *Lorenz system* (Lorenz, 1963), one of the most well-known chaotic systems, involves three coupled ordinary differential equations as follows:

$$\begin{aligned} \dot{x} &= \sigma(y - x) \\ \dot{y} &= -xz + rx - y \\ \dot{z} &= xy - bz \end{aligned} \tag{5.4}$$

These equations have their basis in a Galerkin approximation to the partial differential equations (PDEs) describing thermal convection in the lower atmosphere. In physical terms, the variable x is related to the intensity of the convective motions, y is the temperature difference between the upward and downward convective currents, and z measures the deviation of the vertical temperature profile from a linear one, while σ, r, and b are system parameters. For our experiment, we use the standard parameter settings $\sigma = 16$, $b = 4$, and $r = 45.92$, which are known to give rise to chaos.

In the sequel, the success of the dynamic reconstruction of the Lorenz system from experimental data will be primarily judged by the degree to which it reproduces the following characteristics, that is, the *chaotic* or *dynamic invariants*, of the original system. Our treatment of the material follows Malinetskii (1993) and Tong (1992).

Correlation Dimension D_2 This quantity is a specific version of the *generalized dimension* defined as

$$D_q \triangleq \frac{1}{q-1} \lim_{\epsilon \to 0} \frac{\log \sum_i p_i^q}{\log \epsilon}, \qquad q \in \mathbb{Z}^+ \tag{5.5}$$

Roughly speaking, one covers the set of orbits in the phase portrait of a given chaotic system by cubes of edge ϵ and considers the probability p_i that points of the set lie within the ith such cube. Actually, these generalized dimensions may be applied to arbitrary compact sets so that, for example, when $q = 0$, we may compute the *Kolmogorov capacity* D_C of the familiar one-dimensional Cantor set to be $\log 2/\log 3$ (since 2^n segments of length $1/3^n$ are needed to cover the Cantor set). The importance of the correlation dimension is that it is used in reconstruction theory to determine an appropriate input (autoregressive) order or *embedding dimension* for the reconstructed map.

Lyapunov Spectrum One of the distinguishing characteristics of a chaotic system is its *sensitive dependence on initial conditions*, that is, divergent trajectories resulting from arbitrarily close initial states $\boldsymbol{x}(0)$ and $\boldsymbol{x}'(0)$. In the continuous-time case, set $\boldsymbol{d}(t) \triangleq \boldsymbol{x}'(t) - \boldsymbol{x}(t)$ and define the (first) *Lyapunov exponent* λ

$$\lambda(\boldsymbol{x}(0), \boldsymbol{d}(0)) \triangleq \lim_{t\to\infty} \lim_{\|\boldsymbol{d}(0)\|\to 0} \frac{1}{t} \log \frac{\|\boldsymbol{d}(t)\|}{\|\boldsymbol{d}(0)\|} \tag{5.6}$$

More generally, one may define the d Lyapunov exponents or spectrum $\{\lambda_i\}_{i=1}^{d}$ of a d-dimensional discrete-time dynamical system $\boldsymbol{x}(i+1) = f(\boldsymbol{x}(i))$, $i \in \mathbb{N}$, started from an initial state $\boldsymbol{x}(0)$ by

$$\lambda_i(\boldsymbol{x}(0)) \triangleq \frac{1}{n} \lim_{n\to\infty} \log a_i(n, \boldsymbol{x}(0)) \tag{5.7}$$

where $a_i(n, \boldsymbol{x}(0))$ is the magnitude of the ith largest eigenvalue of the Jacobian matrix $Jf^{(n)}$ of $f^{(n)}$ (the n-fold composition of f) evaluated at $\boldsymbol{x}(0)$. Under suitable conditions (Oseledets, 1968), it can be shown that these exponents are independent of the initial states $\boldsymbol{x}(0)$ and $\boldsymbol{x}'(0)$ when both lie within a neighborhood of a *strange attractor* in the phase space of the chaotic system, where an attractor is considered "strange" if it has at least one positive λ_i. Roughly speaking, for points inside a strange attractor, the *distance* between the trajectories from two infinitesimally close initial states grows with time step i as $\exp(\lambda i) = \exp(\lambda_1 i)$, while the *area* of the triangle defined by the trajectories from three infinitesimally close initial states grows as $\exp((\lambda_1 + \lambda_2)i)$, and so forth. Hence negative exponents are associated with dissipative systems while positive exponents indicate sensitive dependence on initial conditions.

Kaplan–Yorke Dimension D_{KY} The *Kaplan–Yorke dimension* (Kaplan and Yorke, 1978) is defined as

$$D_{\mathrm{KY}} \triangleq M + \frac{\sum_{i=1}^{M} \lambda_i}{|\lambda_{M+1}|} \tag{5.8}$$

where $M \in \mathbb{N}$ is the largest value such that $\sum_{i=1}^{M} \lambda_i > 0$. The significance of this dimension lies in the hypothesis proposed by Kaplan and Yorke (1978) that the Kolmogorov capacity of a strange attractor is equal to its

Kaplan–Yorke dimension, that is, $D_C = D_{KY}$. As noted in Malinetskii (1993), this hypothesis has been verified for a number of continuous-time dynamical systems (up to $d = 3$) and discrete-time dynamical systems (up to $d = 2$). The hypothesis, if proven to hold generally, would provide a fundamental link between the *dynamical* properties (as embodied by its Lyapunov spectrum) of a strange attractor and its *topological* properties (as embodied by its Kolmogorov capacity).

Because these invariants are intrinsic characteristics of the chaotic system and are therefore present in any sample of its realization, they may be considered *global* or *long-term* measures of a given reconstruction. On a *local* or *short-term* basis, an important measure of the quality of a given reconstruction is its (local or short-term) *predictability.* The presence of one or more positive Lyapunov exponents implies that as long as the estimation of an appropriate time-delay embedding from $\mathbf{x}_m(i)$ to $x(i+1)$ is only approximate, that is, the estimated map $\tilde{f}$ does not *exactly* map $\mathbf{x}_m(i)$ to $x(i+1)$ over *all* of the attractors in phase space, there will be an exponential divergence in the evolutions of the original and the reconstructed signals when initialized with the same value. In their studies of the local prediction problem for chaotic systems, Farmer and Sidorowich (1987) provide the approximate recursive relation

$$T \approx \frac{1}{\lambda_{\max}} \log\left(\frac{\sigma_T}{\sigma_0}\right) \tag{5.9}$$

to define the *horizon of predictability* T in terms of the largest Lyapunov exponent of the chaotic process. In the formula, the parameter σ_T is the normalized standard deviation of the prediction error at time T of the RFI prediction process and is defined by

$$\sigma_T^2 = \frac{(1/T)\sum_{j=0}^{T-1}(\tilde{x}(i+j) - x(i+j))^2}{(1/N)\sum_{j=1}^{N}(x(j) - \mu_x)^2} \tag{5.10}$$

where N is the total number of the time series samples and μ_x is the mean value of the samples over the observation period. The parameter σ_0 is the normalized standard deviation of the prediction error at the start of the RFI prediction, that is, at time step i. For noise-free data, σ_T^2 should (in principle) be invariant to the starting point i for the RFI prediction (since the Lyapunov exponents are global characteristics of a chaotic system), hence T should also be independent of the starting point i. In practice, the observed horizon of predictability after reconstruction from simulated data

Table 5.1 Dynamical Invariants for Lorenz System, $\sigma = 16$, $b = 4$, $r = 45.92$

Parameter	D_2	D_{KY}	Lyapunov exponents (nat/s)		
Theoretical value	2.07	2.07	1.50	0.00	−22.50

varies somewhat depending on the initial point, but manifestation of the phenomenon remains clear in results presented below.

For our choice of system parameters, the corresponding theoretical values of the invariants for the system are listed in Table 5.1.

For the purposes of simulation, we numerically solve the Lorenz system equations by fourth-order Runge–Kutta integration with a time step of 0.025 s and the initial conditions $x(0) = y(0) = z(0) = 1.0$ to obtain a 40,000-long signal sequence, of which the first 5000 samples are removed to avoid signal transients (for a sample of the x component of the generated data, see Fig. 5.1). Because the Takens embedding theorem implies that one ought to be able to reconstruct the attractor structure of a chaotic system by observing any one system variable, we concentrate on the x component of the Lorenz system in our studies. In particular, it follows from the Takens embedding theorem that, under suitable conditions, a "smooth" function exists to map $x(i+1)$ from its associated *delayed input vector* $\mathbf{x}_p(i) \stackrel{\Delta}{=} [x(i), x(i-1), \ldots, x(i-p+1)]^T$, that is, the observed system variable is a (deterministic) autoregressive process. Dynamic reconstruction by the estimation of such a map is known as the *delay coordinate embedding* method (Sauer et al., 1991). In this method, the input order p required is determined by the relation

$$p \geq d_E \cdot \tau \tag{5.11}$$

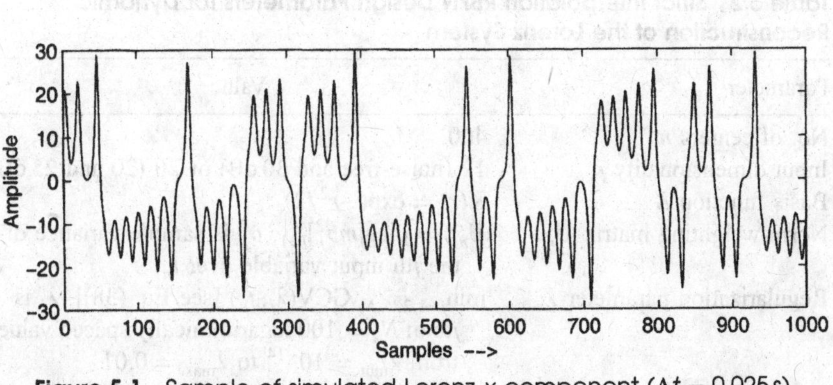

Figure 5.1 Sample of simulated Lorenz x component ($\Delta t = 0.025$ s).

where d_E is the *embedding dimension* and τ is the *characteristic time delay* of the (continuous-time) chaotic system. These two parameters are estimated from simulated time series described above using the *global false nearest neighbors (GFNN)* method (Kennel et al., 1992) and the *mutual information (MI)* method (Fraser and Swinney, 1996; Fraser, 1989; Pineda and Sommerer, 1994), respectively. For the noise-free Lorenz time series, we find that $d_E = 3$ and $\tau = 4$ s, hence for the strict interpolation RBFN, an input dimension of $p = 12$ is sufficient; for the noisy data considered later, different d_E and τ will be necessary, so p will vary. Allowing for this variation in p, the strict interpolation RBFN design parameters common to this and the subsequent experiments are given in Table 5.2. Since the networks in the following studies are implemented with the same software as for the nonlinear speech prediction experiment, details of the parameters can be found in Section 3.7.

With these design parameters, the resultant regularized strict interpolation RBFN is repeatedly trained and operated in the RFI predictive mode beginning at various time steps in the data. Each repetition consists of selecting n-sequential sample data of the time series from, say, time step $i - n + 1$ to i as the training set and then generating an RFI predicted sequence estimating the actual time series at time steps $i + 1$, $i + 2, \ldots, i + M$ for some $M > 0$. Figure 5.2 shows such an RFI predicted sequence beginning at time step $i + 1 = 419$. The predicted sequence follows closely the original sequence up to approximately 300 samples before diverging from it. To visually compare more global structures, we may plot $x(i)$ versus $x(i - 4)$ to produce a two-dimensional phase portrait (the choice of four for the delay is not significant except for yielding a relatively simple phase portrait). The agreement between the

Table 5.2 Strict Interpolation RBFN Design Parameters for Dynamic Reconstruction of the Lorenz System

Parameter	Value
No. of centers n	400
Input dimensionality p	12 (noise-free and 30 dB) or 20 (20 and 25 dB)
Basis function K	$K(r) = \exp(-r^2/2)$
Norm-weighting matrix $\mathbf{U}_n$	$\mathbf{U}_n^{-1} = \mathrm{diag}[m\sigma_j^2]_{j=1}^m$, σ_j^2 is sample variance of the jth input variable over t_n
Regularization parameter λ_n	$\min_{j=1,2,\ldots,N_\lambda} \mathrm{GCV}(\lambda_j, t_n)$ [see Eq. (38)], λ_j is jth of $N_\lambda = 100$ logarithmically spaced values from $\lambda_{\min} = 10^{-14}$ to $\lambda_{\max} = 0.01$

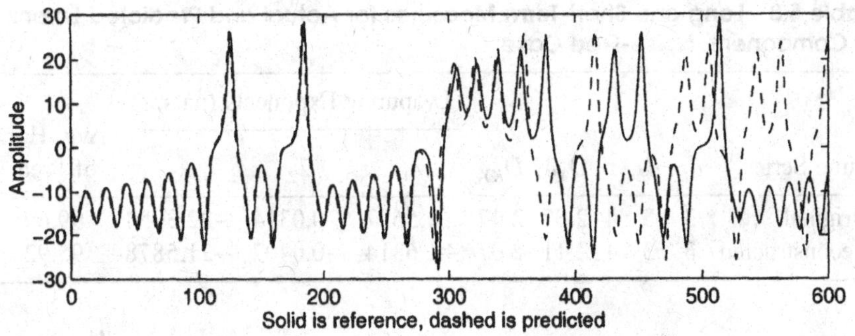

Figure 5.2 Example of RFI predicted sequence for simulated Lorenz x component, noise-free case.

resultant attractor phase portraits of the original and reconstructed systems in Fig. 5.3 are further evidence of a successful reconstruction. The dynamic invariants for the original and reconstructed systems are also compared in Table 5.3. For this purpose, the reconstructed system is used to generate a 35,000-long sample sequence in the RFI predictive mode to be analyzed as follows:

1. The correlation dimension D_2 is estimated by the maximum likelihood-based algorithm described in Schouten et al. (1994). In the table, this estimate is denoted D_{ML}.
2. The Lyapunov spectrum is estimated by the method of Brown et al. (1991). Computationally, this algorithm exploits a recursive QR decomposition of the Jacobian of a function that maps the state variable $x(i)$ to $x(i+n)$, $n > 0$, for different delays n, which, as we have discussed, is central in the definition of Lyapunov exponents.

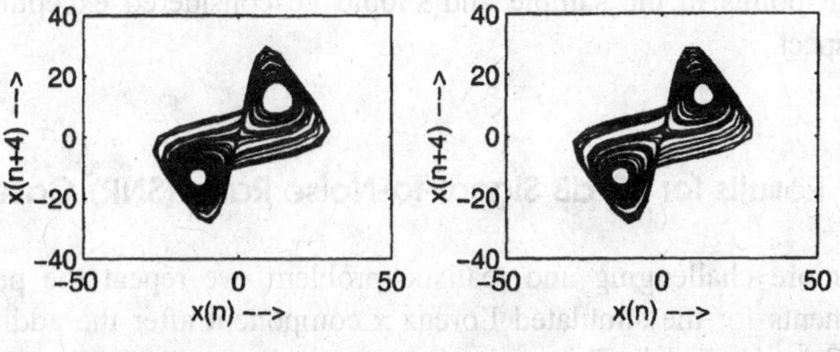

Figure 5.3 Actual (left) and reconstructed (right) attractors projected onto x–y-plane, noise-free case.

Table 5.3 Long and Short-Term Measures for Actual and Predicted Lorenz x Component, Noise-Free Case

Time Series	d_E	d_L	τ	D_{ML}	D_{KY}	Lyapunov Exponents (nats/s) λ_1	λ_2	λ_3	Avg. Hor. of Pred.
Original	3	3	4	2.07	2.07	+1.5697	−0.0314	−22.3054	99.69
Reconstructed	3	3	4	2.11	2.07	+1.6314	−0.0407	−21.5878	95.92

3. The estimated Kaplan–Yorke dimension is computed from the estimated Lyapunov spectrum according to (5.8). Assuming that the corresponding conjecture holds, the estimated Kaplan–Yorke dimension should approximately equal the estimated correlation dimension.

Table 5.3 also lists the estimated *number of local dimensions* or degrees of freedom, as determined by the *local false nearest-neighbors (LFNN)* method (Abarbanel and Kennel, 1993). Because this figure should agree with the actual number of Lyapunov exponents, it is useful as an additional verification of the simulated data quality and subsequent reconstruction.

The close agreement between the estimated chaotic invariants, that is, long-term measures, indicated in Table 5.3 demonstrates that the reconstruction has truly captured the dynamics of the system underlying the original simulated data. The last column of the table also gives the average horizon of predictability as determined from the estimated Lyapunov spectrum by the BBA algorithm of Grassberger (1990). As we noted earlier, the observed horizon of predictability can fluctuate with the initial predicted point; in this regard, Fig. 5.2 shows one of the longest horizons observed when repeating the training and RFI prediction process at different points in the sample and should be considered exceptional in this respect.

5.4.2 Results for 30-dB Signal-to-Noise Ratio (SNR) Case

For a more challenging and realistic problem, we repeat the previous experiments for the simulated Lorenz x component after the addition of white Gaussian noise. This white Gaussian noise is physically generated using a commercially available analog noise generator (NC 1107A-1,

Noise Com, Inc.) coupled with an amplifier and analog/digital (A/D) converter. This device contains a hermetically packaged noise diode that has been burned-in for 168 h and operates in a temperature range of −35 to +100°C to produce white Gaussian noise at power level of +13 dBm over a frequency band between 100 Hz and 100 MHz. Hence the noise sequence added to the original Lorenz data avoids the issues of undesirable correlation and submaximal period that mark noise sequences derived from typical pseudo-random number generators, for example, that in the computational language system MATLAB.

For the most part, the strict interpolation RBFN design parameters in the present noisy case can be the same as for the previous noise-free case, except for the crucial input autoregressive order p. Analysis of the noisy time series by the same methods used in the noise-free case return estimates of $d_E = 5$ and $\tau = 4$, indicating that p should be at least 20 by (5.11). On the other hand, that relation is derived for noise-free chaotic time series, and it is known that using more inputs than necessary can allow the noise to adversely affect the quality of the reconstruction (as farther lagged inputs contribute more "noise" than "information" to the reconstruction). We therefore choose $p = 20$, the minimum permissible input order, for the noisy reconstruction as indicated in Table 5.2. As we shall see, the results obtained suggest that this choice is reasonable.

Using the same training method in the noise-free case, we obtain Fig. 5.4, showing the RFI predicted sequence beginning at time step $i + 1 = 612$. In this case, the RFI predicted sequence successfully tracks the original, that is, *noise-free*, sequence for approximately 230 samples before deviating. We should note, as before, that this RFI sequence is one of the longer ones encountered in the course of repeating the training/prediction cycle with different starting time steps i over the

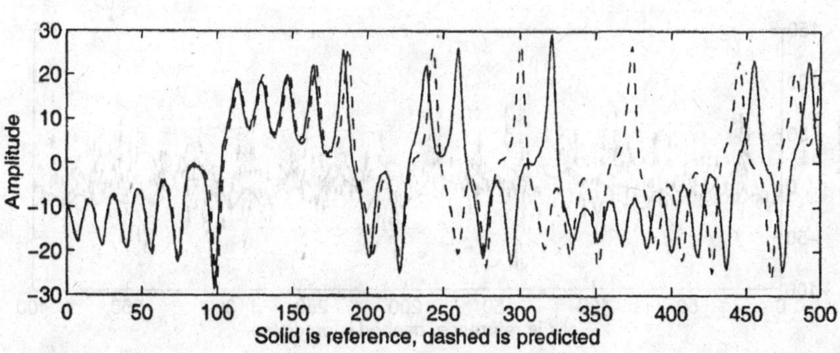

Figure 5.4 Example of RFI predicted sequence for simulated Lorenz x component, 30 dB SNR case.

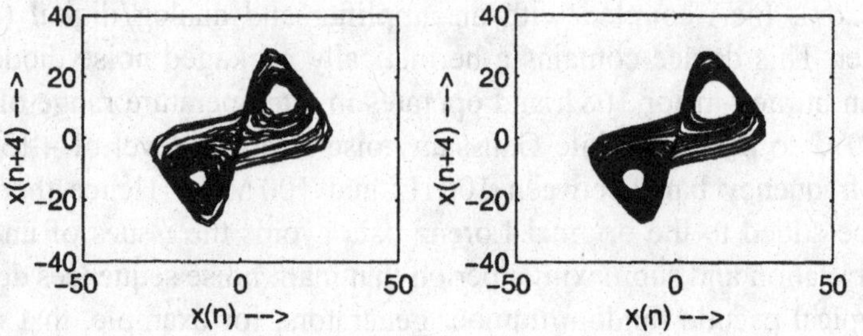

Figure 5.5 Noisy (left) and reconstructed (right) attractors projected onto x–y plane, 30-dB SNR case.

available training data. The corresponding reconstructed attractor can be found in Fig. 5.5, which again, visually, agrees better with the original attractor in Fig. 5.3 than the same attractor for the noisy time series. These results confirm the ability of the regularized strict interpolation RBFN to recover the original dynamics of the system from a noisy observed sequence. We can give an example of how important regularization is to reconstruction from noisy time series by comparing two RFI predicted sequences corresponding to two strict interpolation RBFNs trained according to the parameters of Table 5.2 with the same 25 dB SNR sample sequence. Specifically, Fig. 5.6 displays the RFI predicted sequence again beginning from time step $i + 1 = 612$ produced by a nonregularized, that is, $\lambda = 0$, strict interpolation RBFN while Fig. 5.7 shows the same sequence when the strict interpolation RBFN is regularized through the usual GCV-based method listed in Table 5.2. The inability of the nonregularized strict interpolation RBFN to discern the true underlying dynamics from the noisy data is particularly graphic in this example. The dynamic invariants and the horizon of predictability for this

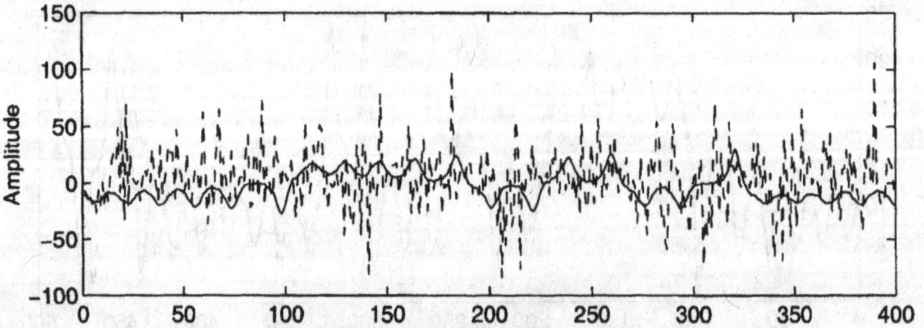

Figure 5.6 Example of RFI predicted sequence for simulated Lorenz x component, 25-dB case with no regularization.

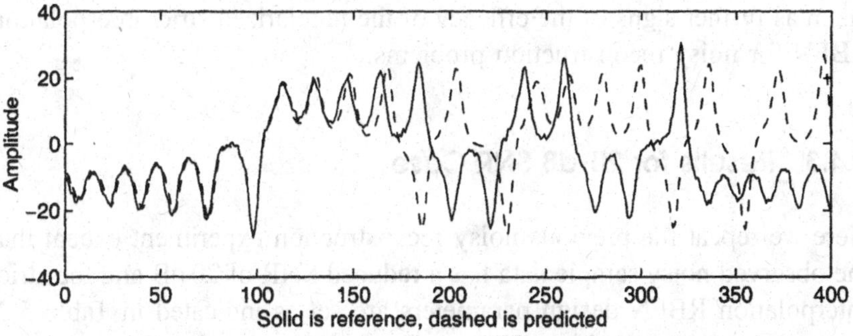

Figure 5.7 Example of RFI predicted sequence for simulated Lorenz *x* component, 25-dB case with regularization.

30-dB SNR case are given in Table 5.4, where the figures are computed in the same fashion as in the noise-free case. Note that even at the relative high 30-dB SNR level, the added noise is sufficient to mislead greatly the estimation algorithms for the chaotic invariants previously described, for example, the first Lyapunov exponent for the noise-corrupted signal is estimated to be 10.43 compared to 1.57 for the original (noise-free) signal. The added noise also allows the Lyapunov spectrum estimation algorithm to detect a spurious fourth Lyapunov exponent of −45.00, which is not theoretically present in the original Lorenz system. In comparison, the regularized strict interpolation RBFN reconstruction "denoises" sufficiently the noisy signal both to yield an RFI predicted sequence whose estimated first Lyapunov exponent of 2.17 is much closer to that of the original signal and to avoid the introduction of an extraneous fourth Lyapunov exponent. As for the average horizon of predictability, here again the reconstructed signal gives an estimate of 72 samples, which show less deterioration from the original horizon of 99 samples than the 15 samples estimated from the noisy signal. These results too may be

Table 5.4 Long-and Short-Term Measures for Noisy and Predicted Lorenz *x* Component, 30-dB SNR Case

Time Series	d_E	τ	D_{ML}	D_{KY}	Lyapunov Exponents (nats/s) λ_1	λ_2	λ_3	λ_4	Avg. Hor. of pred.
30-dB Signal	4	4	2.25	2.83	+10.4287	+0.9483	−13.5706	−45.0007	15.01
Reconstructed	3	4	2.02	2.09	+2.1667	−0.4793	−18.1266	–	72.22

taken as further signs of the efficacy of the regularized strict interpolation RBFN for noisy reconstruction problems.

5.4.3 Results for 20-dB SNR Case

Here we repeat the previous noisy reconstruction experiment except that the observed noisy sample data has a reduced SNR of 20 dB and the strict interpolation RBFN design parameters are set as indicated in Table 5.2. One of the better tracking RFI predicted sequences from the resultant regularized strict interpolation RBFN estimate beginning at time step $i+1 = 422$ can be seen in Fig. 5.8. Despite the increased noise level, the reconstructed system can still yield an RFI predicted sequence that corroborates with the corresponding sequence from the original time series up to approximately 150 samples. The smoothing effect of the regularized strict interpolation RBFN reconstruction is clear in Fig. 5.9, where the associated reconstructed attractor is shown beside the same attractor derived from the noisy time series. Qualitatively, the reconstructed attractor still exhibits a definite similarity to the original, noise-free attractor in Fig. 5.3. The same can be said for the long-term dynamical invariants estimated for the reconstructed system as listed in Table 5.5. As in the 30-dB SNR case, we observe that when analyzed with the standard chaotic parameter estimation algorithms, the noisy time series returns values that are far from the true values of the underlying noise-free series. In particular, the presence of noise adds to the divergence of the local trajectories, resulting in a grossly overestimated first Lyapunov exponent of 16.60 for the noisy time series versus 1.57 for the original (noise-free) time series. Again, we also see that the added noise results in

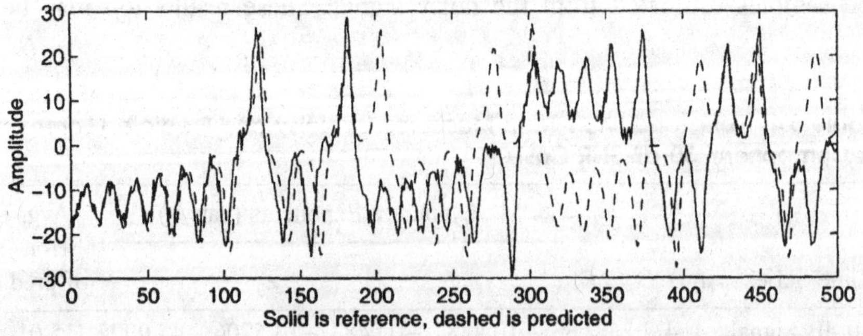

Figure 5.8 Example of RFI predicted sequence for simulated Lorenz x component, 20-dB SNR case.

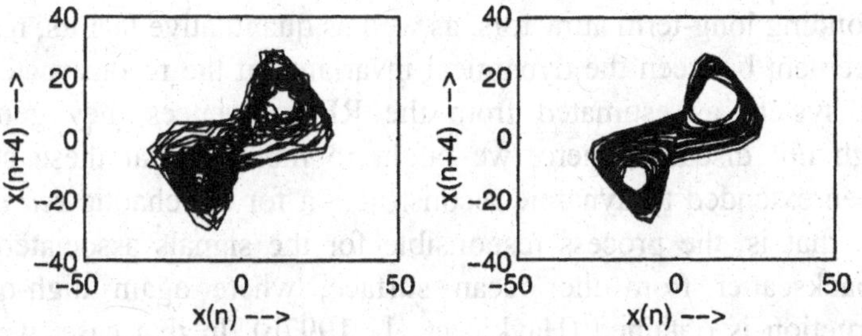

Figure 5.9 Noisy (left) and reconstructed (right) attractors projected onto x–y plane, 20-dB SNR case.

an estimated Lyapunov spectrum with extraneous exponents, except that in this case there is an additional minimum exponent $\lambda_5 = -46.7312$ (not listed in Table 5.5 because of space constraints). The reconstructed system, however, suffers from neither of these problems, although the accuracy of the estimated invariants is understandably decreased at this relative low SNR. We may therefore conclude that dynamic reconstruction using the regularized strict interpolation RBFN remains a viable method even in the presence of a moderate amount of noise.

5.5 SUMMARY AND DISCUSSION

The results of the simulations demonstrate that the regularized strict interpolation RBFN (with regularization parameter chosen via an a.o. procedure) is an effective tool for the reconstruction of chaotic dynamical systems from its observed time series, even in the presence of moderate amounts of noise. Here, the quality or success of the reconstruction is judged according to both qualitative factors, namely, the agreement (up to the horizon of predictability) between short-term RFI predicted sequences compared to the original sequence starting at the same point and their

Table 5.5 Long- and Short-Term Measures for Noisy and Predicted Lorenz x Component, 20-dB SNR Case

Time Series	d_E	τ	D_{ML}	D_{KY}	Lyapunov Exponents (nats/s) λ_1	λ_2	λ_3	λ_4	Avg Hor. of pred.
20 dB signal	5	4	3.15	4.15	+16.5982	+8.4169	−2.1425	−15.6714	9.42
Reconstructed	3	4	2.09	2.12	+2.6379	−0.2902	−19.8489	–	59.32

corresponding long-term attractors, as well as quantitative factors, namely, the agreement between the dynamical invariants of the reconstructed and original system as estimated from the RFI sequences they generate. Although not discussed here, we ought to mention that these studies have been extended to dynamic reconstruction for the chaotic *sea clutter* process, that is, the process responsible for the signals associated with radar backscatter from the ocean surface, where again high-quality reconstruction is obtained (Haykin et al., 1997b). In that case, we have an example of dynamic reconstruction from noisy observations of a real-life process for which there is not (as yet) a definitive mathematical model. It is in such contexts that the flexible nonparametrics behind the regularized strict interpolation RBFN method can arguably find greatest application.

Despite the success of the regularized strict interpolation RBFN method as a tool for dynamic reconstruction in the experiments conducted, many open questions remain as raised in the discussion of Section 5.2. Foremost among these questions may be those concerning the stability of the reconstructed system in the RFI predictive mode. While this stability was not an issue for the simulated Lorenz data, a stable reconstruction of the considerably more complex (as measured by the correlation coefficient D_2) sea clutter process did require careful selection of the strict interpolation RBFN design parameters through some trial-and-error procedure. A straightforward approach to the stability issue is to constrain the strict interpolation RBFN parameters, specifically the weights and the regularization parameter, so that the resultant map, when used recurrently in the RFI predictive mode, is a (globally) contractive map [for example, see Steck (1992)]. Unfortunately, this approach cannot allow reproduction of more complex attractor structures such as the quasiperiodic orbits seen in the Lorenz examples. Clearly more refined techniques are necessary to achieve stable, high quality reconstruction from observational data, noisy or otherwise.

6 DISCUSSION

In this concluding chapter, we do two things:

1. Review the results presented in this book from a broader perspective to place them in the context of current and past developments.
2. Discuss the positive and negative aspects of our approach along with its possible extension.

Since the resurgence of interest in artificial neural networks (ANNs), the radial basis function network (RBFN) has stood alongside the multilayer perceptron (MLP) as one of the paradigms of choice in a large variety of applications. Yet the theoretical properties of the actual choices made in popular RBFN design procedures remain largely unknown, save for some general results concerning the density of certain restricted classes of RBFNs. We therefore posed the question: What practical RBFN design procedures are theoretically justified? In this book, for the term *theoretically justified*, we confined our attention primarily to (global) mean-square consistency, as is common in engineering applications—other measures of network performance are of course possible, as we shall indicate below.

6.1 RELATIONSHIP BETWEEN RADIAL BASIS FUNCTION NETWORKS AND KERNEL REGRESSION ESTIMATES

We began our study by examining the theoretical justification for current ANN and kernel regression estimate (KRE) approaches to RBFN design, that is, selection of network parameters. In this respect, the breadth and depth of the empirical risk minimization (ERM) theory can rightly be considered a breakthrough compared with previous *ad hoc* attempts. While the basic ERM approach can be provably consistent, in practice some sort of architectural constraint on the network is necessary to avoid the undesirable extremes of underfitting (excessive estimator bias) and overfitting (excessive estimator variance) for finite training sets. The structural risk minimization (SRM) method addresses this problem in a principled manner by generalizing the quadratic regularizers used in regularized RBFNs to impose an *a priori* structure on the candidate solution space; but this leaves open the question of how such a suitable structure should be determined. On an operational level, when nonquadratic cost functionals are specified, the SRM method typically results in a difficult optimization problem, except in special circumstances. Also the minimization of the guaranteed risk central to the SRM method is not necessarily the type of optimality desired in applications.

On the other hand, it appears as if the ANN community is not fully aware of how the large, relatively mature body of theory surrounding the KRE can be relevant to the RBFN design problem. Perhaps this lack of interest stems, in part, from the notion within some ANN circles that ANNs constitute a "special" and separate class from traditional statistical estimators. Nonetheless, the similarity between the KRE, particularly in its most prevalent form, the Nadaraya–Watson regression estimate (NWRE), and the strict interpolation RBFN is obvious, leading to some of the studies mentioned in the Introduction. These studies, however theoretically intriguing, do not relate to regularized RBFNs as they are typically constructed and hence are of limited practical value. The question thus remains open: Does there exist another path to consistent RBFN design other than through the ERM/SRM theory?

The affirmative answer proposed in this book is based on a combination of elements from KRE and penalized least-squares (PLS)/spline smoothing theory and practice. It turns out, not surprisingly, that the key is the presence of the additive quadratic regularizer with parameter λ. In conjunction with the original least-squares cost function, the resultant simple algebraic equations for the optimum weights in the strict interpolation case give a direct link to the NWRE through λ, albeit under some restrictions. It is interesting to note that the link takes the form of constructive asymptotic approximation theorems in sup-norm as well as the usual L_2-norm, suggesting that both modes of consistency could carry over from the NWRE to the regularized strict interpolation RBFN. Although this intuition can be proven correct, there is little to be gained if a regularized strict interpolation RBFN is designed to approximate asymptotically a given NWRE in either mode; after all, if it is NWRE behavior one desires, one could just use the NWRE rather than its strict interpolation RBFN approximation.

Instead, what makes the approximation theorems nontrivial is the knowledge from PLS and spline smoothing theory concerning the optimal selection of the regularization parameter λ with respect to the risk or mean-square fitting error (MSFE). Using one of the asymptotically optimal (a.o.) parameter selection methods for λ, we can ensure that the MSFE of a regularized strict interpolation RBFN is (asymptotically) minimum over all other choices for λ, including the MSFE for the corresponding NWRE-equivalent strict interpolation RBFN. This result can be useful in itself, for example, in the case where the training inputs represent important operating points, but often we are more interested in the global mean-square error (MSE). For this purpose, we introduce a theorem relating the MSFE to the global MSE under conditions compatible with those assumed for the NWRE approximation theorems. Together with the approximation results, we can conclude that the regularized strict interpolation RBFN is globally MS consistent whenever the corresponding NWRE is and where the regularization parameter has been chosen to be asymptotically optimal. The implied prescription for provably (MS) consistent RBFN design is therefore:

- From KRE theory, start with an MS consistent NWRE design. For the RBFN, choose as the basis function the kernel/bandwidth combination of the NWRE. For simplicity, we can set one basis function for each training input datum as in the NWRE, that is, use a strict interpolation RBFN. Note, however, that this choice is not necessary if NWRE approximation theorems can be obtained for nonstrict interpolation RBFNs. Even limiting ourselves to nonstrict interpolation methods based on selecting a subset of the available centers for basis functions, a more sophisticated analysis than in the strict case would be necessary to establish corresponding constructive approximation results. Once the centers of the RBFN are selected, such nonstrict interpolation RBFNs have the practical advantages of lower training time and run-time complexity, as may be expected. Of course, the optimal centre selection problem can itself present a considerable theoretical and computational challenge.

- From PLS/spline smoothing theory, choose λ via a suitable asymptotically optimal parameter selection procedure. By "suitable," we mean one that is known to be asymptotically optimal under the given network operating conditions, for example, homoskedastic or heteroskedastic noise. As we saw in the nonlinear state estimation experiments in Chapter 4, an asymptotically optimal parameter selection cannot only be used to select effectively the regularization parameter λ but also the input scaling parameters that weight the relative contributions of each (covariance) normalized coordinate in the network input vector. The resultant input scaling factors then indicate which input variables are of lesser importance to the network output, leading to a form of input variable selection or dimensionality reduction. Given the importance of this topic in data analysis, its continued study is warranted.

It can certainly be argued that the route to provably consistent RBFN design proposed in the above is neither the most direct nor the most elegant from a theoretical standpoint; in particular, it should well be possible to produce a proof for the consistency of the regularized strict interpolation RBFN without recourse to approximating the NWRE (although the choice of regularization parameter would still require justification). That said, there are some definite merits to our approach:

1. We can exploit the substantial knowledge that exists concerning the NWRE in the statistical regression context. Because of its simplicity, the analysis of the NWRE has yielded a comprehensive understanding of its properties under various important scenarios such as mixing processes and some forms of nonstationarity. In fact, as we have previously discussed, the formal analyses of the properties of the (regularized strict interpolation) RBFN that are available have usually made the (rather strong) assumption of i.i.d. training data. The relaxation of this assumption can be considered one of the desirable consequences to our approach.
2. We have considerable theoretical and practical experience with asymptotically optimal parameter selection methods, the latter of which is embodied by the availability of efficient computational methods in some special cases and the numerous case studies under

different conditions. This significant corpus is not something to be ignored in applications.

6.2 LIMITATIONS OF THE APPROACH TAKEN IN THIS BOOK

There are naturally some limitations that arise from our synthetic approach to a justifiable RBFN design procedure; we point some of them out here to suggest avenues for further research:

1. The strict interpolation condition, that is, one basis function per training input datum, can result in impractically large networks for significantly sized problems, as well as a loss of numerical stability during training that can yield low-quality estimates. In principle, choosing a high degree of smoothing, that is, λ sufficiently large, should be able to alleviate the stability problem but this choice could run counter to the MSFE optimality condition that λ decrease to zero as the number of training data (but not necessarily basis functions) goes to infinity. As we mentioned before, it should be possible to include the nonstrict interpolation case in our synthetic approach once the appropriate approximation theorems are in place.
2. Because most cross-validation (CV) theory centers around asymptotic optimality with respect to the squared-error loss and its expected value, the risk or MSFE, our approach can currently only demonstrate MS consistency. A direct analysis of the regularized strict interpolation RBFN would not be subject to this restriction. We should point out that some research on the properties of the generalized CV (GCV) estimate for the optimal λ under other loss functions has been performed. For example, Wahba (1990) reports of situations in spline smoothing where the GCV estimate for the optimal λ also (approximately) minimizes the integrated squared-error, that is, R_2^* in Eq. (3) in the Introduction with uniform input distribution. Even these preliminary results, however, indicate that the properties (more precisely, the rate of decay) of the minimizing λ for the different loss functions vary with the exact circumstances of the problem at hand. It ought to be possible to apply other cross-validatory methods that select the regularization parameter to minimize (asymptotically) other loss functions such as the mean absolute value, resulting in a "hybrid" least-squares estimator. We could then prove that the regularized strict interpolation RBFN with

regularization parameter selected according to such a CV method is consistent in the mode specified by its corresponding loss function.

3. The (asymptotic) rate to MS consistency of a regularized strict interpolation RBFN (with asymptotically optimal selection of λ) in our approach is lower bounded by, that is, is at least as fast as, the rate for the corresponding NWRE, and upper bounded by the rate at which the MSFE converges to the global MSE, a quantity that we estimated crudely in the proof of Theorem 2 to be $\mathcal{O}(n^{-1/d})$ under only uniform boundedness and Lipschitzian assumptions on the functions concerned. We expect that significantly improved rates can be obtained by assuming additional regularity, that is, smoothness, for the functions in the theorem. These assumptions are reasonable for RBFNs since smooth basis functions such as the Gaussian are commonly used and since it is known that the underlying function must be assumed smooth if estimation is to be possible in high-dimensional spaces.

Of course, there are approaches to RBFN design from entirely different perspectives than the one we have chosen to take in this book. At this point, we therefore briefly discuss some of these approaches in order to provide an expanded view of RBFNs and their place within the broad spectrum of ANN methods and techniques.

6.3 DESIGN OF RBF NETWORKS USING DETERMINISTIC ANNEALING

Deterministic annealing is an optimization procedure that offers three important attributes (Ross, 1998; Ross et al., 1992):

1. The ability to alleviate the problem of local minima of a cost function with adjustable parameters.
2. Applicability to a variety of different network structures.
3. Ability to overcome the vanishing gradient problem by minimizing the cost function even when its gradient with respect to the adjustable parameters may vanish.

Like *simulated annealing* due to Kirkpatrick et al. (1983), deterministic annealing is rooted in statistical mechanics. Both of these annealing procedures rely on the analogy between the behavior of a physical system with many degrees of freedom in thermal equilibrium at a series of finite temperatures (as experienced in statistical mechanics) and the problem of finding the minimum of a cost function with respect to a set of adjustable parameters. Simulated annealing is, however, a stochastic *relaxation* procedure in that it wanders over the *energy surface* while incremental adjustments are made on the average, and for the procedure to converge to a global minimum, the annealing *schedule* has to be chosen carefully. In other words, the rate at which the temperature is successively lowered must be carefully chosen, a requirement that renders the use of simulated annealing unrealistic for many practical applications. *Deterministic* annealing overcomes this limitation by using expectation (or averaging) in place of stochastic relaxation. Consequently, deterministic annealing is able to escape local minima without the slow annealing process that characterizes simulated annealing. There is, however, no guarantee that deterministic annealing will converge to a global minimum.

With regards to RBFNs, Rao et al. (1997) apply the deterministic annealing (DA) procedure to RBF classifier design. Zheng et al. (2000) have extended the application of DA to nonlinear regression using normalized RBFNs. The reader will recall that in the Introduction normalized RBFNs are discussed as one obvious way of linking the NWRE with RBFNs. Here DA is used to train the centers of the RBFN. As a measure of clustering quality, the following MSE is defined as the cost function to be minimized:

$$J(\mathcal{X}, \mathcal{C}) \triangleq \sum_{i,j} P(\boldsymbol{x}_i^{(j)}) \|\boldsymbol{x}_i^{(j)} - \boldsymbol{c}_j\|^2 \tag{6.1}$$

where $\mathcal{X} \triangleq \{\boldsymbol{x}_i\}$ is the input sample set, $\mathcal{X}_j \triangleq \{\boldsymbol{x}_i^{(j)}\}$ is the input subset pertaining to the jth class, $\mathcal{C} \triangleq \{\boldsymbol{c}_j\}$ is the codebook (that is, the set of clustering centers), and $P(\boldsymbol{x}_i^{(j)})$ is the sample probability pertaining to the jth class. The cost function $J(\mathcal{X}, \mathcal{C})$ is non-differentiable, which negates the direct use of gradient-based minimization methods. To overcome this difficulty, differentiable functions are used to approximate the cost function $J(\mathcal{X}, \mathcal{C})$. Specifically, the following function is introduced:

$$\|f_{\boldsymbol{x}}(\mathcal{C})\|_\infty = \max_j(-\|\boldsymbol{x} - \boldsymbol{c}_j\|^2) = -\min_j(-\|\boldsymbol{x} - \boldsymbol{c}_j\|^2) \tag{6.2}$$

Accordingly, the cost function $J(\mathcal{X}, \mathcal{C})$ is reformulated as

$$J(\mathcal{X}, \mathcal{C}) = -\sum_i P(\boldsymbol{x}_i) \| f_{\boldsymbol{x}_i}(\mathcal{C})\|_\infty \tag{6.3}$$

Typically, the input samples are independent, in which case we may set $P(\mathbf{x}_i) \stackrel{\Delta}{=} 1/k$. Correspondingly, the minimization of $J(\mathcal{X}, \mathcal{C})$ is equivalent to

$$\min_{\mathcal{C}} \min_{j}(\|\boldsymbol{x} - \boldsymbol{c}_j\|^2) \tag{6.4}$$

The Lagrangian function for this optimization problem is

$$L(\mathcal{C}, \boldsymbol{u}) = \sum_{j} u_j \|\boldsymbol{x} - \boldsymbol{c}_j\|^2 \tag{6.5}$$

where the u_j are the Lagrange multipliers. The Lagrangian function so defined is a convex combination of $\|\boldsymbol{x} - \boldsymbol{c}_j\|^2$. Hence, we may recast the optimization problem of (6.4) as that of finding the distribution that minimizes the objective function of (6.5) subject to a specific level of randomness defined by the Shannon entropy

$$H(\boldsymbol{u}) = -\sum_{j} u_j \log u_j \tag{6.6}$$

With these definitions at hand, the stage is set for the application of DA as described in (Ross, 1998). In particular, the optimization of (6.4) is restated as one of minimizing the Lagrangian

$$\min_{\boldsymbol{u}} L_T(\mathcal{C}, \boldsymbol{u}), \qquad L_T(\mathcal{C}, \boldsymbol{u}) \stackrel{\Delta}{=} L(\mathcal{C}, \boldsymbol{u}) - TH(\boldsymbol{u}) \tag{6.7}$$

where T is the Lagrangian multiplier that plays the role of temperature in a statistical mechanics context. The optimization thus proceeds by first selecting a large value of T and then reducing it gradually as the minimization of (6.7) is repeated.

Using the optimization procedure summarized herein, Zheng et al. (2000) present a system identification experiment that compares the following ANN architectures:

- Multilayer perceptron and RBFN both trained using the back-propagation algorithm.
- RBFN trained using the DA procedure.

The results presented therein show that the RBFN trained using deterministic annealing not only improves upon the generalization ability of the backpropagation-trained architectures but also has reduced the training time. These reported results should be suggestive of the increases in performance possible if the RBFN centers are admitted as network parameters to be optimized during the training process. On the other hand, with respect to the central theme of this book, Zheng et al. (2000) do not analyze the consistency of their proposed DA procedure for RBFN center selection, leaving its theoretical properties in this regard unknown.

6.4 SEMI-PARAMETRIC MACHINE LEARNING

Regularized RBFNs, as discussed in this book, provide a powerful nonparametric tool for solving nonlinear regression and classification problems using empirical data. The success of this approach to machine learning follows the formulation of the problem in terms of a cost function with a built-in regularization constraint.

There are, however, many practical situations in signal processing, for example, modeling of radar clutter, where linear estimators are known to provide reasonable performance, but may be improved through the addition of a nonlinear capability. In such situations, a purely nonparametric approach may not provide the best trade-off between complexity and generality. Rather, we may be able to achieve a better solution by adopting a *semi-parametric approach* whereby an unregularized but known parametric component is added to the model. In doing so, we can combine the practical simplicity of a linear model with the generality of a non-parametric model. Indeed, these remarks are borne out in the speech prediction application of Chapter 3, where linearly combining the outputs of several nonlinear speech predictors formed by regularized RBFNs along with the usual autoregressive signal inputs yielded a nontrivial performance improvement compared with prediction using the autoregressive signal inputs alone. While this hybrid linear-nonlinear structure can be justified (on the basis of MMSE prediction theory) independently of the considerations behind the regularized RBFN, the approach we describe below integrates the design of both the linear and nonlinear components from the very onset.

To be specific, the usual cost functional of a regularized RBFN in Eq. (6.11) in the Introduction is modified so that the resultant cost functional to be minimized is

$$\begin{aligned} H(f, \psi) &\triangleq H_1(f, \psi) + \lambda H_2 f, \\ H_1(f, \psi) &\triangleq \sum_{i=1}^{N} (y_i - f(\boldsymbol{x}_i, \boldsymbol{w}) - \psi(\boldsymbol{x}_i, \boldsymbol{a}))^2, \\ H_2 f &\triangleq \Phi(f) \end{aligned} \tag{6.8}$$

where, as usual, $\{(\boldsymbol{x}_i, y_i)\}_{i=1}^N$ constitutes the training set, f and ψ represent the nonparametric and parametric parts of the model (respectively), λ is the regularization parameter, and Φ is a penalty functional that constrains the space of possible solutions in accordance with some form of *a priori* knowledge. The nonparametric and parametric parts of the model are defined in terms of the weight vectors $\boldsymbol{w}$ and $\boldsymbol{a}$, respectively.

Mattera et al. (2000) argue for the effectiveness of such semi-parametric modeling when the parametric part represents a linear component with long memory while the nonparametric part represents a nonlinear component with (relatively) short memory. In that work, the *support vector machine (SVM)* is used for the nonlinear component. Briefly, the SVM is a linearly weighted network of the form described in Eq. (1) in the Introduction, for example the RBFN, designed by an algorithm that is an approximate method for SRM. For our purposes, the SVM is the function that minimizes the regularized cost functional in Eq. (11) in the Introduction with the *ϵ-sensitive* error functional $H_1 f$, where

$$H_1 f \stackrel{\Delta}{=} \sum_{i=1}^{N} |y_i - f(x_i)|_{\epsilon+} \tag{6.9}$$

and $|\bullet|_{\epsilon+}$ is the ϵ-sensitive L_1 distance defined by

$$|\bullet|_{\epsilon+} \stackrel{\Delta}{=} \begin{cases} 0, & |\bullet| < \epsilon \\ \bullet - \epsilon, & \text{otherwise} \end{cases} \tag{6.10}$$

It can be shown (Mattera et al., 2000) that the optimality conditions satisfied at the minimum of Eq. (11) in the Introduction with $H_1 f$ as in Eq. (6.9) lead to a *constrained* quadratic programming problem whose solution, while theoretically well understood, is computationally involved and typically requires iterative optimization methods such as gradient or coordinate descent. When a linear parametric component is introduced into (6.9) according to (6.8), an additional equality constraint is placed on the optimum direction vector $\Delta\tilde{\boldsymbol{a}}(k)$ at iteration k for the iterative updating of the estimated optimum linear component weight vector $\tilde{\boldsymbol{a}}(k)$ via $\tilde{\boldsymbol{a}}(k+1) = \tilde{\boldsymbol{a}}(k) + \mu_k \Delta\tilde{\boldsymbol{a}}(k)$, where μ_k is the update step size. In situations where the number of weights in the parametric component is large, fulfilling this equality constraint can be problematic, leading Mattera et al. (2000) to develop an alternative method based on the augmented Lagrangian that results in a *sequence* of minimization problems involving only the nonparametric/nonlinear component. For the regularized RBFN, the addition of a linear component such as that embodied by ψ in (6.8) is handled easily due to the continuity of the cost functional $H_1 f$ in the linear parameters $\boldsymbol{a}$. Specifically, if $\psi(\boldsymbol{x}, \boldsymbol{a}) \stackrel{\Delta}{=} \boldsymbol{x}^\top \boldsymbol{a} : \mathbb{R}^d \mapsto \mathbb{R}$ is a linear function[1] and $H_2 f$ is as defined in Eq. (11) in the Introduction, then we find that the estimated optimal weights $\tilde{\boldsymbol{w}}$ and $\tilde{\boldsymbol{a}}$ of the nonlinear and linear components, respectively, satisfy the joint system of equations

$$(\boldsymbol{G} + \lambda \boldsymbol{I})\tilde{\boldsymbol{w}} = \boldsymbol{y} - \boldsymbol{X}\tilde{\boldsymbol{a}} \tag{6.11}$$

$$(\boldsymbol{X}^\top \boldsymbol{X})\tilde{\boldsymbol{a}} = \boldsymbol{X}^\top (\boldsymbol{y} - \boldsymbol{G}\tilde{\boldsymbol{w}}) \tag{6.12}$$

(where $\boldsymbol{X}$ is an appropriately ordered row-wise matrix of the $\boldsymbol{x}_i^\top$) that can be solved straightforwardly. Although not discussed herein, the theory on regularized RBFNs described in this book can be extended naturally to incorporate a "rough" model represented by the parametric component ψ in (6.8).

6.5 CONTINUOUS LEARNING IN A REGULARIZED MANNER

On another practical level, our approach has not addressed the issue denoted in the ANN literature as *continuous learning*. In the nonparametric function estimation context that we have adopted, the term denotes the process of continually adapting and refining a function estimate in response to new data that describe changes in the unknown target function. State-space models of the form studied in Chapters 4 and 5 but with *time-variant* state and observation functions (which we did not explore in any detail) would be natural candidates for continuous learning problems. Indeed, even the partial and full update algorithms for strict interpolation RBFN in the speech prediction task of Chapter 3, like the linear adaptive recursive least-squares (RLS) algorithm that inspired them, may be considered elementary expressions of continuous learning. Despite the success achieved with these techniques, the *ad hoc* and relatively unsophicated nature of their approaches to nonstationarity, for example, resetting the function estimate on large prediction errors (for the partial update algorithm) and recursive reestimation (for the full update algorithm), merely point to the need for a more basic reexamination of the problem of *tracking* time-varying phenomena. The pipelined recurrent

[1]For example, the $\boldsymbol{x}_i$ can be the autoregressive input vectors described in Chapter 3, in which case we may think of $\boldsymbol{a}$ as containing the tap values of a FIR filter.

MLPs of Haykin and Li (1995), Baltersee and Chambers (1998), and Mandic and Baltersee (1999) mentioned in Chapter 3 represent one accepted way of continuously estimating time-varying relationships from observational data. Here we shall briefly introduce the method developed by Ormoneit (1999) for continuous learning. What distinguishes this approach from others is the principled use of regularization, a constant theme in this work, to introduce *tracking* into the variational cost function of Eq. (11) of the Introduction whose minimizer defines the (time-varying) function estimate. Specifically, Eq. (11) is replaced by the more general time-varying form

$$HF^n \triangleq H_1F^n + \lambda H_2F^n \qquad H_1F^n \triangleq \sum_{i=1}^{n}(y_i - f(\boldsymbol{x}_i;\, \boldsymbol{\theta}_i))^2$$
$$H_2F^n \triangleq \sum_{i=2}^{n} \|\boldsymbol{\theta}_i - \boldsymbol{\theta}_{i-1}\|^2 \tag{6.13}$$

where F^n is a *sequence* of n functions $\{f(\bullet;\, \boldsymbol{\theta}_i)\}_{i=1}^{n}$ and f is a common network architecture or structure parameterized by a (real-valued) vector $\boldsymbol{\theta} \in \mathbb{R}^d$. The estimated function sequence is then defined as the one satisfying

$$\widetilde{F^n} = \arg \min_{F^n \in \mathcal{F}^n(f)} HF^n \tag{6.14}$$

where $\mathcal{F}^n(f)$ denotes the space of all function sequences of length n and each function is a parameterized version of the common function f. Although similar in form to the original cost function, the cost function proposed by Ormoneit (1999) for continuous learning contains some significant differences:

1. The candidate solution space over which the minimizer is sought is a space of function *sequences* rather than functions themselves. Note, however, for fixed n, that restricting the component functions in the sequence to different parameterizations of a common function allows us to recast the problem as equivalent to one of static function estimation with a candidate solution space of functions mapping $\mathbb{R}^{nd}$ to $\mathbb{R}$.

2. The regularizing term H_2F^n penalizes the smoothness of the candidate function *sequence* (equivalently, the corresponding function parameter sequence) over *time*, that is, as new training data arrive. In the usual static function estimation scenario, the regularizing term $H_2 f$ penalizes the smoothness of the candidate *function* over its *domain*.

3. Whereas the usual static function estimation scenario uses a training data *set* to specify the unknown function behavior, the continuous learning scenario uses a training data *sequence* to do the same. The regularized solution to the static function estimation problem is invariant to permutations of the training data set, while the *order* of the training data is important to the specification of the regularized continuous learning problem. In the latter case, it is intuitive that permuting the training data input–output pairs in the training data sequence can result in a significantly different estimated regularized solution function sequence. For example, let the input dimensionality $d = 1$ and consider a sequence of training data $S_1^n \triangleq \{(x_i, y_i)\}_{i=1}^n$ generated by $y = f(x) = 1/x$ over $x_i = i$, $i = 1, 2, \ldots, n$. Presented in the "natural" order, we should think that S_1^n would generate the same solution function ($\tilde{f} = f$) for the static function estimation scenario as it would for the continuous learning scenario $[\widetilde{F^n} = F^{n,*} = \{f, f, \ldots, f\}$ (n times)]. On the other hand, had we used the permuted training sequence $S_2^n \triangleq \{(x_1, y_1), (x_n, y_n), (x_2, y_2), (x_{n-1}, y_{n-1}), \ldots\}$, it is quite conceivable that, depending on how well the candidate function set $\{f(\bullet; \boldsymbol{\theta}) : \theta \in \mathbb{R}^d\}$ can model the inverse function, we may well end up with a function sequence very different from the optimal sequence $F^{n,*}$.

One may think of the (Ormoneit, 1999) regularized cost function for continuous learning as seeking the most smoothly (time-varying) function-sequence which explains the observed training data sequence. As one might suspect, the minimization of Eq. (6.13) is somewhat more involved than that required in the static function estimation scenario. Roughly speaking, the approach adopted by Ormoneit (1999) is to consider the penalty term $H_2 F^n$ to be the negative logarithm of an appropriately defined Gaussian prior probability density $p_{\boldsymbol{\Theta}^n}$ over the space of length n parameter sequences. Similarly, the error term $H_1 F^n$ can be considered the negative logarithm of an appropriately defined Gaussian conditional probability density $p_{S^n|\boldsymbol{\Theta}^n}$ of the observed data sequence S^n given the parameter sequence $\boldsymbol{\Theta}^n$. With these interpretations, the minimization of (6.13) is then equivalent to seeking the maximum *a posteriori* (MAP) estimate $\widetilde{\boldsymbol{\theta}^n}$ of the conditional probability density $p_{\boldsymbol{\Theta}^n|S^n}$. Ormoneit (1999) discusses how the iterated extended Kalman filter (IEKF) is naturally suited to solving recursively for this MAP estimate via iterations comprising a forward *filtering* pass and a backward *smoothing* pass. When the common function f is nonlinear in its parameters $\boldsymbol{\theta}$ (as is normally the case), multiple iterations of the IEKF are generally required to obtain a stable MAP estimate of the optimal network parameter sequence.

To demonstrate the real-world efficacy of the technique, Ormoneit (1999) applies it to the problem of predicting the prices and hedge ratios of financial derivatives.[2] While it is beyond the scope of this brief discussion to delve into the details of pricing derivative assets such as options, what is relevant is that, for options on the German stock exchange index DAX, an MLP trained using the regularized continuous learning algorithm performed comparably to the widely accepted benchmark *Black–Scholes model*[3] while exceeding the performance of both online backpropagation and EKF-trained MLPs in terms of the relative improvement in predicting option prices and the mean hedging error. The results also show that the MSE of the proposed algorithm's option pricing error, while initially worse than the reference Black–Scholes model, can be improved by more than an order of magnitude by exploiting *a priori* knowledge of the theoretical boundary behaviour of option prices near expiry.[4] Although these results demonstrate clearly that the regularized continuous learning algorithm of (Ormoneit, 1999) is a powerful one, this capability is not without its disadvantages:

1. Unlike a block-mode technique such as the regularized strict interpolation RBFN and its dynamic counterparts, the inherently sequential and iterative nature of the regularized continuous learning algorithm can be computationally complex. As the training sequence of length n is progressively passed through the regularized continuous learning algorithm, older data are reused by the algorithm more and more often, so that the first datum $i = 1$ is processed by the algorithm n times, the $i = 2$ datum is processed by the algorithm $n - 1$ times, and so forth, leading to a minimum $\mathcal{O}(n^2)$ computational complexity, where each IEKF iteration of the algorithm consists of a forward/backward pass. All of this assumes the best-

[2]For a detailed exposition on the application of ANNs to problems involving financial derivatives, see Hutchison (1994), Hutchinson et al. (1994). For general background on financial derivatives, see (Hull, 1993).

[3]The *Black–Scholes* model for option pricing assumes that the price of the underlying asset being optioned is a particular type of stochastic process (an *Ito process*) with drift and diffusion parameters that are dependent on the risk-free rate of return and the asset's volatility, respectively. Thus the Black–Scholes formula is an explicitly *parametric* model [see Black and Scholes (1973)].

[4]This result corroborates with the well-known principle that nonparametric models such as ANNs and KREs can benefit significantly for judiciously chosen *a priori* knowledge concerning their unknown target functions.

case where only one IEKF iteration is required for convergence per training datum introduced. In practice, multiple iterations per introduced training datum may be required for convergence. As in the recursive regularized strict interpolation RBFN algorithms of Chapter 3, online processing is practical only if measures such as truncating the training sequence to the most recent n training data are taken.

2. An issue not addressed in the proposed regularized continuous learning algorithm of Ormoneit (1999) is the estimation of the optimal regularization parameter λ. Indeed, Tables 2 and 3 in (Ormoneit, 1999) show how the performance of the proposed algorithm is strongly dependent on a good choice of regularization parameter λ, for example, a four-to-one ratio between the MSEs corresponding to the "worst" and "best" choices of λ considered. Since this optimal λ must be chosen on the basis of *sequences* of functions (or, equivalently, parameters) each of whose computation can be itself quite demanding, evaluation of the optimal λ by a CV type of proxy may be considered too burdensome. In this case, suboptimal cruder techniques such as the trial-and-error evaluation of λ logarithmically over some nominal range used in the results for the proposed algorithm may be employed.

In spite of these shortcomings, the theoretical and practical potential of this line of thinking regarding the problem of continuous learning is evidently promising and worthy of further research. It is perhaps most fitting that we end here on a consonant note: There are obviously many other possible developments for regularized RBFNs (and nonparametric estimators in general) that may be pursued, for example, nonasymptotic estimation error bounds. We hope, however, that the discussions in this chapter circumscribe clearly the main boundaries of the work reported in this book and offer an interesting glimpse at what could be done. In the end, one could say that this book bears two messages: To the ANN community, the KRE/NWRE theory has much to offer as a guide for justifiable RBFN design under a wide variety of conditions; To the kernel regression community, the extension provided by the presence of the regularization parameter is nontrivial, computationally feasible, and well supported in theory and practice from the PLS/spline smoothing field. We have aimed to bring exactly these messages together, and if it succeeds in influencing the thinking of either community, then the effort was well spent.

APPENDIX OF NOTES TO THE TEXT

A.1 NOTES FOR INTRODUCTION

1. The *Lebesgue measure* in $\mathbb{R}^d$ can be defined as the measure that assigns to any set of the form $\{\boldsymbol{x} \in \mathbb{R}^d : a_i \leq x_i < b_i, i = 1, 2, \ldots, d\}$, for some a_i, b_i satisfying $-\infty < a_i < b_i < \infty$, the number $\prod_{i=1}^{d}(b_i - a_i)$. For example, the Lebesgue measure represents the length, area, and volume of a set in $d = 1$-, 2-, or 3-dimensional Euclidean space, respectively.
2. We restrict our attention to the case where P is a probability measure defined over $\mathbb{R}^d$. Then for a P-measurable sequence of RVs $\{f_n\}$, we say that:

   (a) $f_n \overset{n\to\infty}{\to} 0$ i.p.–P if $P\{\boldsymbol{x} \in \mathbb{R}^d : |f_n(\boldsymbol{x})| > 0\} \overset{n\to\infty}{\to} 0$, that is, the indicated convergence to zero occurs over a set of points whose probability (under P) approaches one.

   (b) $f_n \overset{n\to\infty}{\to} 0$ a.s.–P if $P\{\boldsymbol{x} \in \mathbb{R}^d : \lim_{n\to\infty} f_n(\boldsymbol{x}) = 0\} = 1$, that is, the indicated convergence to zero occurs at every point possible under the distribution P.

   (c) $f_n \overset{n\to\infty}{\to} 0$ in $L_p(D, P)$ if $(\int_D |f_n|^p dP)^{1/p} \overset{n\to\infty}{\to} 0$, for example, for $D = \mathbb{R}^d$ and $p = 2$, we have the familiar mean-square convergence.
3. A function $g : \mathbb{R}^d \to \mathbb{R}$ is *Borel measurable* if the set $\{\boldsymbol{x} \in \mathbb{R}^d : g(\boldsymbol{x}) \in (a, b)\}$ is measurable for each bounded open interval $(a, b) \in \mathbb{R}$.

4. A *Hilbert space* is an inner product space that is complete in the norm induced by the inner product. For example, the familiar space of Lebesgue square-integrable functions defined on some (measure) space is a Hilbert space [for more details, see, for example, (Halmos, 1957)].
5. A *pseudo-differential* operator P is defined by duality with the algebraic action of its Fourier transform $\hat{P} = \hat{P}(\boldsymbol{s})$, where $\boldsymbol{s}$ is the d-dimensional transform variable corresponding to the (input) variable $\boldsymbol{x}$. In other words, the operator equation

$$(Pf)(\boldsymbol{x}) = g(\boldsymbol{x}) \tag{A.1}$$

is synonymous with the multiplicative equation

$$\hat{P}(\boldsymbol{s})\,\hat{f}(\boldsymbol{s}) = \hat{g}(\boldsymbol{s}) \tag{A.2}$$

where $\hat{\bullet}$ denotes the Fourier transform. Hence a pseudo-differential operator P is well-defined if its Fourier transform $\hat{P}(\boldsymbol{s})$ is a convergent series in $\boldsymbol{s}$. For example, in the 1-D case, if P is the pseudo-differential operator

$$P = \sum_{n=0}^{\infty} \frac{(-1)^n}{n!} \frac{d^n}{dx^n} \tag{A.3}$$

then the Fourier transform which defines its action is given by

$$\hat{P}(s) = \sum_{n=0}^{\infty} \frac{(-1)^n s^n}{n!} = \exp(-s) \tag{A.4}$$

6. The *Schwartz theory of tempered distributions* considers the action of continuous linear functionals defined over a (Hilbert) space of "infinitely smooth" functions $\mathcal{S}$. A function $f \in \mathcal{S}$ if f and all its partial derivatives decrease to zero faster than any power of $\|\boldsymbol{x}\|^{-1}$ as $\|\boldsymbol{x}\| \to \infty$. More precisely, $f \in \mathcal{S}$ if, for some constants C_{mk} (independent of $\boldsymbol{x}$), $\|\boldsymbol{x}\|^m |D^k f(\boldsymbol{x})| \leq C_{mk}$ for all $m, k \in \mathbb{N}$ and $\boldsymbol{x} \in \mathbb{R}^d$, where D^k is a kth order partial derivative. For more details, see, for example, (Zemanian, 1987) or (Friedlander, 1982).
7. In fact, Light (1992) shows that when the basis functions G_i are derived from a positive-definite kernel, the resultant interpolation matrix $\boldsymbol{G}$ is positive definite (for any distinct choices of centers).

Recall that a dyadic function $G : \mathbb{R}^d \times \mathbb{R}^d \to \mathbb{R}$ is positive definite if for all n and $a_1, a_2, \ldots, a_n \in R$, $\boldsymbol{x}_1, \boldsymbol{x}_2, \ldots, \boldsymbol{x}_n \in \mathbb{R}^d$, we have $\sum_{i,j=1}^{n} a_i a_j G(\boldsymbol{x}_i, \boldsymbol{x}_j) > 0$.

8. Following the notation of footnote A.2, the *Borel–Cantelli* lemma states that if

$$\sum_{n=1}^{\infty} P\{\boldsymbol{x} \in \mathbb{R}^d : |f_n(\boldsymbol{x})| > 0\} < \infty \tag{A.5}$$

then $f_n \overset{n\to\infty}{\to} 0$ a.s.–P. We may view the lemma as providing a method of proving the a.s. convergence of a sequence of RVs from their basic large deviation behaviour.

9. In the *method of sieves*, the target function f is assumed to lie in a (possibly infinite dimensional) space $\mathcal{F}$ whose members admit arbitrary approximation by a sequence of parametric models with an increasing number of parameters.

10. A *Radon* measure μ over $\mathbb{R}^d$ is a Borel measure for which (a) every subset of $\mathbb{R}^d$ is contained in a Borel set of equal μ measure and (b) every compact subset K of $\mathbb{R}^d$ has finite μ-measure. For more details and examples, see Girosi and Anzellotti (1995).

11. A *finite polynomial spline* is a finite, piecewise polynomial function. A *sufficiently kinky polynomial spline* is a finite polynomial spline in which either (a) one of the highest order polynomial segments adjoins a lower order polynomial segment or (b) all of the polynomial segments have the same order and two of them have different leading coefficients. For examples and further discussion, see Stinchombe and White (1990).

12. An RKHS is a Hilbert space $H(K; T)$ of (real-valued) functions defined over a domain T with a *reproducing kernel* K satisfying (a) $\forall t \in T$, $K(\bullet, t) \in H(K; T)$ and (b) $f(t) = \langle f, K(\bullet, t)\rangle_{K,T}$, where $\langle \bullet, \bullet \rangle_{K,T}$ is the inner product in $H(K; T)$. A simple example of an RKHS is the space $\mathbb{R}^d$ of (real) d vectors with $T \triangleq \{1, 2, \ldots, d\}$ and inner product $\langle \boldsymbol{x}, \boldsymbol{y}\rangle_{K,T} \triangleq \boldsymbol{x}^\top \boldsymbol{K}^{-1} \boldsymbol{y}$, where $\boldsymbol{x}$, $\boldsymbol{y} \in \mathbb{R}^d$ and $\boldsymbol{K}$ is a symmetric positive definite matrix. In this case, $\boldsymbol{K}$ is the reproducing kernel since for $j \in T$, $\boldsymbol{k}_j$, the jth column (vector) of $\boldsymbol{K}$, satisfies $\langle \boldsymbol{x}, \boldsymbol{k}_j\rangle_{K,T} \triangleq \boldsymbol{x}^\top \boldsymbol{K}^{-1} \boldsymbol{k}_j = \boldsymbol{x}^\top \boldsymbol{e}_j = x_j$, where $\boldsymbol{x} \triangleq [x_1, x_2, \ldots, x_d]$ and $\boldsymbol{e}_j$ is the jth basis vector in $\mathbb{R}^d$.

13. Let $\{X_i\}_{i\in\mathcal{I}}$ (where $\mathcal{I}$ is some index set) be a sequence of RVs, that is, a stochastic process. Let F_n^m be the σ field generated by $\{X_i\}_{n\le i\le m}$, that is, F_n^m is the space of all events involving the RVs

$\{X_i\}_{n \le i \le m}$. Then (a) $\{X_i\}_{i \in \mathcal{I}}$ is a ϕ *mixing* process if $\phi_k \triangleq \sup_{i \in \mathcal{I}} \sup_{A \in F^n_{-\infty}; P(A)>0, B \in F^{\infty}_{n+k}} |P(B|A) - P(B)|$ satisfies $\lim_{k \to \infty} \phi_k = 0$. (b) $\{X_i\}_{i \in \mathcal{I}}$ is a α *mixing* process if $\alpha_k \triangleq \sup_{i \in \mathcal{I}} \sup_{A \in F^n_{-\infty}, B \in F^{\infty}_{n+k}} |P(A \cap B) - P(A)P(B)|$ satisfies $\lim_{k \to \infty} \alpha_k = 0$. In other words, these mixing conditions describe processes for which any pair of events involving their past and future become asymptotically independent as the (index) separation between the events grows. From the definition of conditional probability, it is clear that ϕ mixing implies α mixing but the converse is not necessarily true. It is intuitive that the familiar class of pth order *autoregressive* [$AR(p)$] processes is ϕ mixing. For more details, see Douhkan (1994) and Bradley (1986).

14. The PLS theory of Golitschek and Schumaker (1990) considers linear basis function expansions $\tilde{f}_n \triangleq \sum_{i=1}^{n} w_i G_i$ for the approximation of an unknown function f from noisy training data where the measurement noise sequence $\{\epsilon_i\}$ is i.i.d. zero-mean constant variance. The linear weights $\boldsymbol{w}$ are chosen to minimize a risk or cost function of the form (17) except that the penalty term $||\sqrt{\boldsymbol{G}}\boldsymbol{\omega}||^2$ is replaced with the more general expression $\boldsymbol{\omega}^{\top}\boldsymbol{E}\boldsymbol{\omega}$ where $\boldsymbol{E}$ is a symmetric nonnegative-definite matrix.

A.2 NOTES FOR CHAPTER 1

1. The *Fröbenius norm* of a square matrix $\boldsymbol{A}$ is defined as the square root of the sum of squared magnitudes of the matrix elements.
2. Taking the point of view that a stochastic process $\{X(i)\}$ is an indexed collection of random variables (RVs), we can compute the marginal density p_i of $X(i)$ for each i. If $p_i \equiv p$ for all i, then $\{X(i)\}$ is said to have a common marginal density p.
3. If D is a compact set, define D_ϵ as follows: $A \in D_\epsilon$ if A is compact and for all $\boldsymbol{x} \in A$, there is a $\boldsymbol{y} \in D$ such that $||\boldsymbol{x} - \boldsymbol{y}|| \le \epsilon$. Thus D_ϵ is the collection of compact sets whose elements are not further than ϵ away from some element of D.
4. The *Lebesgue-dominated convergence theorem* states that if $\{f_n\}$ is a sequence of integrable functions converging a.e. to some function f and $|f_n| < g$ a.e. for some fixed integrable function g, then f is also integrable and $\lim \int f_n d\mu = \int \lim f_n \, d\mu = \int f \, d\mu$. The theorem

therefore gives conditions for the usual interchange of limit and integral.

5. A function f is Lipschitz over D if for all $\boldsymbol{x}$ and all $\boldsymbol{y} \in D$, there exists a constant $K > 0$ such that $|f(\boldsymbol{x}) - f(\boldsymbol{y})| \leq K\|\boldsymbol{x} - \boldsymbol{y}\|$.
6. Given $p \geq 1$ and $k \in \{0, 1, 2, \ldots, \}$, a *Sobolev space* $W_p^k(\Omega)$ over some open set $\Omega \subset \mathbb{R}^d$ can be defined as $W_p^k(\Omega) \triangleq L_p(\Omega) \cap \{f : D^\alpha f \in L_p(\Omega), |\alpha| < k\}$ where D^α is a partial derivative of order α (Ziemer, 1989). In such spaces, it can be shown that the MS convergence (1.8) implies the corresponding pointwise convergence (1.6).
7. Following the notation that we have established, a point z is in the Voronoi cell $V_{z_n}(z(i))$ centered about $z(i)$ if z is closer to $z(i)$ than any other point in z_n, that is, $V_{z_n}(z(i)) \triangleq \{z \in \mathbb{R}^d : \|z - z(i)\| \leq \|z - z(j)\|, \forall j = 1, 2, \ldots n\}$.
8. The classical *ridge regression* problem can be stated as: given an input matrix $\boldsymbol{G} \in \mathbb{R}^{N \times N}$ and a vector of desired outputs $\boldsymbol{y} \in \mathbb{R}^N$ related via $\boldsymbol{y} = \boldsymbol{G}\boldsymbol{w} + \epsilon$, where $\epsilon = [\epsilon_i]_{i=1}^N$ is a vector of Gaussian zero-mean RVs with common variance σ, estimate the regression vector $\boldsymbol{w}$ as $\boldsymbol{w} = \arg\min_{\boldsymbol{\omega} \in \mathbb{R}^N} (\|\boldsymbol{y} - \boldsymbol{G}\boldsymbol{\omega}\|^2 + \lambda\|\boldsymbol{\omega}\|^2)$ (Wahba, 1990).

A.3 NOTES FOR CHAPTER 2

1. In fact, Theorem 25.9 of Devroye et al. (1996) states that for the kernel regression estimate (KRE) of posterior class probabilities with the rectangular kernel (see item 1a in Section 1.2), there exist some probability distributions for which the sequence of mean-squared error minimizing bandwidth estimates result in a probability of classification error diverging to infinity with respect to the true minimum probability of classification error. In other words, the minimum mean-square error approximating KRE for posterior class probabilities can (if used in approximate Bayes decision rules) give arbitrarily poor classification performance compared to the true Bayes optimal classifier.
2. Theorem 29.2 in Section 29.3 of Devroye et al. (1996) gives the conditions for the strong Bayes risk consistency of approximate Bayes rules generated from LSFI. More specifically, Theorem 29.8 in Section 29.6 shows how LSFI with linear basis function expansions whose weighting coefficients grow sufficiently slowly with the number of basis functions can lead to (strong) universal Bayes risk consistency.

3. The *cardinality* of a set A is the size of the set as determined by the existence of a one-to-one mapping to another reference set, for example, $\mathbb{N}$ (the natural numbers) or $\mathbb{R}$ (the real numbers). For example, any collection of n objects has a cardinality of n by a 1–1 correspondence with the subset $\{1, 2, \ldots, n\} \in \mathbb{N}$ and A is called a *finite* set. Any infinite set A whose elements may be enumerated or ordered as $\{a_1, a_2, a_3, \ldots,\}$ by a one-to-one correspondence with the whole of $\mathbb{N}$ is called *countable* and has a cardinality denoted by the symbol $\aleph_0$ (aleph null).
4. *Jensen's inequality* implies that for any convex function $g : \mathbb{R} \to \mathbb{R}$, $g(\mathbb{E}[X]) \leq \mathbb{E}[g(X)]$ where X is any real-valued RV with finite mean. Recall that a function g is convex if $g(\alpha x + (1-\alpha)y) < \alpha g(x) + (1-\alpha)g(y)$ for all $x, y \in \mathbb{R}$ and $0 \leq \alpha \leq 1$.

A.4 NOTES FOR CHAPTER 4

1. The *filtration* $\mathcal{F}^X(t)$ of a stochastic process $\{X(t) : t \in \mathcal{T}\}$ is defined as the σ field generated by $\{X(s) : s \leq t,\ s, t \in \mathcal{T}\}$, that is, $\mathcal{F}^X(t)$ is the space of all events involving the process X up to time t. Specifically for our case, one may think of $\mathcal{F}^Y(t)$ as encoding the history of the process B_s up to time t.
2. For example, the quantity being estimated in the SDE-based methods is often a suitable version $u(t, \boldsymbol{x})$ of the conditional state density, in which case *pathwise* convergence may be stated as $\sup_{t \in \mathcal{T}} \int_D |\tilde{u}_n(t, \boldsymbol{x}) - u(t, \boldsymbol{x})|^2 \, d\boldsymbol{x} \overset{n\to\infty}{\to} 0$ for almost all sample paths, where $\tilde{u}_n(t, \boldsymbol{x})$ is the estimate of $u(t, \boldsymbol{x})$ over the domain D. Sun and Glowinski (1993) prove such a result for their method of *operator splitting* to solve the *Zakai filtering equation* (related to the original dynamical system via a linear partial differential equation) and also provide an estimate (under certain conditions) of the rate of convergence. On the other hand, *strong* (resp. *weak*) L_2 convergence for the same problem is the condition $\mathbb{E}[\int_{t \in \mathcal{T}} \int_D |\tilde{u}_n(t, \boldsymbol{x}) - u(t, \boldsymbol{x})|^2 \, d\boldsymbol{x} \, dt] \overset{n\to\infty}{\to} 0$ a.s. (resp. i.p.).

REFERENCES

Abarbanel, H. D. I. (1996). *Analysis of observed chaotic data*. New York: Springer.

Abarbanel, H. D. I., and Kennel, M. B. (1993). Local false nearest neighbors and dynamical dimensions from observed chaotic data. *Phys. Rev. E*, *47*, 3057–68.

Andrews, D. W. K. (1991). Asymptotic optimality of generalized C_L, cross-validation, and generalized cross-validation in regression with heteroskedastic errors. *J. Econometrics*, *47*, 359–77.

Aronszajn, N. (1950). Theory of reproducing kernel Hilbert spaces. *Trans. Amer. Math. Soc.*, *68*, 337–404.

Auestad, B., and Tjøstheim, D. (1990). Identification of nonlinear time series: first order characterization and order determination. *Biometrika*, *77*(4), 669–87.

Baker, G. L., and Gollub, J. P. (1996). *Chaotic dynamics: an introduction*. Cambridge, UK: Cambridge University Press.

Baltersee, J., and Chambers, J. A. (1998). Nonlinear adaptive prediction of speech with a pipelined recurrent neural network. *IEEE Trans. Signal Processing*, *46*(8), 2207–2216.

Barnard, E. (1992). Comments on 'Bayes statistical behavior and valid generalization of pattern classifying neural networks'. *IEEE Trans. Neural Networks*, *3*(6), 1026–7.

Black, F., and Scholes, M. (1973). Pricing options and corporate liabilities. *J. Political Economy*, *81*, 637–654.

Bosq, D. (1973). Sur l'estimation de la densité d'un processus stationnaire et mélangeant. *Comptes Rendus d'Academie Sciences, Series A, Paris*, *277*, 535–538.

Bosq. D. (1975). Inégalité de Bernstein pour les processus stationnaires et mélangeant. Applications. *Comptes Rendus d'Academie Sciences, Series A, Paris*, *281*, 1095–1098.

Bosq. D. (1996). *Nonparametric statistics for stochastic processes* (Vol. 110). New York: Springer-Verlag.

Bradley, R. C. (1986). Basic properties of strong mixing conditions. In E.

Eberlein and M. S. Taqqu (Eds.), *Dependence in probability and statistics* (Vol. 11, p. 165–192). Birkhauš.

Broomhead, D. S., and Lowe, D. (1988). Multivariable functional interpolation and adaptive networks. *Complex Systems*, *2*, 321–335.

Brown, R., Bryant, P., and Abarbanel, H. D. I. (1991). Computing the Lyapunov exponents of a dynamical systems from observed time series. *Phys. Rev. A*, *43*, 2787–2806.

Cacoullos, T. (1966). Estimation of a multivariate density. *Ann. Inst. Stat. Math. (Tokyo)*, *18*(2), 179–189.

Casdagli, M. (1989). Nonlinear prediction of chaotic time series. *Physica D*, *35*, 335–356.

Corradi, V., and White, H. (1995). Regularized neural networks: Some convergence rate results. *Neural Computation*, *7*, 1225–1244.

Craven, P., and Wahba, G. (1979). Smoothing noisy data with spline functions: estimating the correct degree of smoothing by the method of generalized cross-validation. *Numer. Math.*, *31*, 377–403.

Devroye, L. (1981). On the almost everywhere convergence of nonparametric regression function estimates. *Ann. Statist.* *9*(6), 1310–1319.

Devroye, L., and Györfi, L. (1985). *Nonparametric density estimation, the L_1 view.* New York: Wiley.

Devroye, L., Györfi, L., and Lugosi, G. (1996). *A probabilistic theory of pattern recognition*. New York: Springer.

Douhkan, P. (1994). *Mixing: properties and examples* (Vol. 85). New York: Springer-Verlag.

Epanechnikov, V. A. (1969). Nonparametric estimation of a multidimensional probability density. *Theor. Probab. Appl.*, *14*, 153–158.

Eubank, R. L. (1988). *Spline smoothing and nonparametric regression* (Vol. 90). New York: Marcel Dekker.

Farmer, J. D., and Sidorowich, J. J. (1987). Predicting chaotic time series. *Physics Review Letters*, *59*(8), 845–848.

Földes, A. (1974). Density estimation for dependent sample. *Studia Scientarium Mathematicarum Hugarica*, *9*, 443–452.

Fraser, A. M. (1989). Information and entropy in strange attractors. *IEEE Trans. Infor. Theory*, *35*, 245–262.

Fraser, A. M., and Swinney, H. L. (1986). Independent co-ordinates for strange attractors from mutual information. *Phys. Rev. A*, *33*(2), 1134–1140.

Friedlander, F. G. (1982). *Introduction to the theory of distributions*. Cambridge, UK: Cambridge University Press.

Gallant, A. (1987). *Nonlinear statistical models*. New York, NY: Wiley.

Geman, S., Bienenstock, E., and Doursat, R. (1992). Neural networks and the bias/variance dilemma. *Neural Computation*, *4*, 1–58.

Girosi, F., and Anzellotti, G. (1995). *Convergence rates of approximation by translates* (A.I. Memo No. 1288). Cambridge, MA: MIT.

Glick, N. (1972). Sample-based classification procedures derived from density estimators. *J. Amer. Statist. Assoc.*, *67*, 116–122.

Golitschek, M. von, and Schumaker, L. L. (1990). Data fitting by penalized least squares. In J. C. Mason and M. G. Cox (Eds.), *Algorithms for approximation II* (p. 210–227). London: Chapman and Hall.

Grassberger, P. (1990). An optimised box-assisted algorithm for fractal dimension. *Phys. Lett. A*, *148*, 63–68.

Gray, R. M. (1987). *Probability, random processes, and ergodic properties*. New York: Springer-Verlag.

Grenander, U. (1981). *Abstract inference*. New York: Wiley.

Guyon, I., Vapnik, V., Boser, B., Bottou, L., and Solla, S. A. (1992). Structural risk minimization for character recognition. In D. Touretzky (Ed.), *Advances in Neural Information Processing Systems 4* (p. 105–119). San Mateo, CA: Morgan Kaufmann.

Györfi, L. (1978). On the rate of convergence of nearest neighbour rules. *IEEE Trans. Info. Theory* (24), 509–512.

Györfi, L., Härdle, W., Sarda, P., and Vieu, P. (1989). *Nonparametric curve estimation from time series* (Vol. 60). Heidelberg, Germany: Springer-Verlag.

Hager, W. W. (1989). Updating the inverse of a matrix. *SIAM Review*, *31*(2), 221–239.

Halmos, P. R. (1957). *Introduction to Hilbert space*. New York: Chelsea Publishing Company.

Hampshire, J. B., II, and Perlmutter, B. (1990). Equivalence proofs for multi-layer perceptron classifiers and the Bayes discriminant function. In *Proc. of the Connectionist Models Summer School* (p. 159–172). San Mateo, CA: Morgan Kaufman.

Hand, D. (1982). *Kernel discriminant analysis* (Vol. 2). Chichester, UK: Research Studies Press.

Härdle, W. (1990). *Applied nonparametric regression* (Vol. 19). Cambridge, UK: Cambridge University Press.

Haykin, S. (1996). *Adaptive filter theory* (3rd ed.) Englewood Cliffs, NJ: Prentice Hall.

Haykin, S., and Li, L. (1995). Non-linear adaptive prediction of nonstationary signals. *IEEE Trans. Signal Processing*, *43*, 526–535.

Haykin, S., Sayed, A. H., Zeidler, J., Yee, P., and Wei, P. (1997a). Adaptive tracking of linear time-variant systems by extended RLS algorithms. *IEEE Trans. Signal Processing*, *45*(5), 1118–1128.

Haykin, S., Puthusserypady, S., and Yee, P. (1997b). *Dynamic reconstruction of a chaotic process using regularized RBF networks* (CRL Report No. 353).

McMaster University, Hamilton, Canada: Communications Research Laboratory.

Haykin, S., Yee, P., and Derbez, E. (1997c). Optimum nonlinear filtering. *IEEE Trans. Signal Processing*, *45*(11), 2774–2786.

He, X., and Lapedes, A. (1993). Successive approximation radial basis function networks for nonlinear modeling and prediction. In *Proc. IJCNN* (Vol. 2, p. 1997–2000). Nagoya, Japan.

Hull, J. C. (1993). *Options, futures, and other derivative securities* (2nd ed.). Englewood Cliffs, NJ: Prentice-Hall.

Hutchinson, J. M. (1994). *A radial basis function approach to financial time series analysis*. Ph.D. thesis, Dept. of EECS, MIT.

Hutchinson, J. M., Lo, A., and Poggio, T. (1994). A nonparametric approach to pricing and hedging derivative securities via learning networks. *J. Finance*, *49*, 771–818.

Kadirkamanathan, V., and Kadirkamanathan, M. (1996). Recursive estimation of dynamic modular RBF networks. In *Advances in neural information processing systems* (Vol. 8, p. 239–45). San Mateo, CA: Morgan Kaufman.

Kadirkamanathan, V., and Niranjan, M. (1993). A function estimation approach to sequential learning with neural networks. *Neural Computation*, *5*, 954–975.

Kalman, R. E. (1960). A new approach to linear filtering and prediction problems. *Trans. ASME, J. Basic Eng.*, *82*, 35–45.

Kanaya, F., and Miyake, S. (1991). Bayes statistical behavior and valid generalization of pattern classifying neural networks. *IEEE Trans. Neural Networks*, *2*(4), 471–475.

Kaplan, K., and Yorke, J. (1978). Functional differential equations and the approximatioin of fixed points. In *Lecture notes in mathematics* (Vol. 730, p. 228–237). New York: Springer-Verlag.

Kennel, M. B., Brown, R., and Abarbanel, H. D. I. (1992). Determining embedding dimension for phase-space reconstruction using a geometrical construction. *Phys. Rev. E*, *45*, 3403–3411.

Kirkpatrick, S., Gelatt, C. D., and Vecchi, M. P. (1983). Optimization by simulated annealing. *Science*, *220*(3), 671–680.

Krzyżak, A., Linder, T., and Lugosi, G. (1996). Nonparametric estimation and classification using radial basis functions. *IEEE Trans. Neural Networks*, 7(2), 475–487.

Li, K. C. (1985). From Stein's unbiased risk estimates to the method of generalized cross validation. *Ann. Statist.*, *13*(4), 1352–1377.

Li, K. C. (1987). Asymptotic optimality for C_p, C_L, cross-validation and generalized cross-validation: discrete index set. *Ann. Statist.*, *15*(3), 958–975.

Light, W. A. (1992). Some aspects of radial basis function approximation. In S. P. Singh (Ed.), *Approximation theory, spline functions and applications* (Vol. 356, p. 163–190). Dordrecht, Netherlands: Kluwer.

Lo, J. T. (1994). Synthetic approach to optimal filtering. *IEEE Trans. Neural Networks*, *5*(5), 803–811.

Lo, J. T. (1995). *Neural network approach to optimal filtering* (Tech. Rep. No. RL-TR-94-197). Griffiss Air Force Base, NY: Rome Laboratory, Air Force Material Command.

Lorenz, E. N. (1963). Deterministic non-periodic flows. *J. Atmos. Sciences*, *20*, 130–141.

Lowe, D. (1995). On the use of nonlocal and non positive definite basis functions in radial basis function networks. In *Proc. IEE ANN* (p. 206–211). Cambridge, UK.

Lowe, D., and McLachlan, A. (1995). Modelling of nonstationary processes using radial basis function networks. In *Proc. IEE ANN* (p. 300–35). Cambridge, UK.

Malinetskii, G. G. (1993). Synergetics, predictability, and deterministic chaos. In Y. A. Kratsov (Ed.), *Limits of predictability* (Vol. 60, p. 75–141). New York: Springer-Verlag.

Mandic, D. P., and Baltersee, J. A. (1999). Toward an optimal PRNN-based nonlinear predictor. *IEEE Trans. Neural Networks*, *10*(6), 1435–1442.

Maria, F. D. de, and Figueiras-Vidal, A. R. (1995). Nonlinear prediction for speech coding using radial basis functions. In *Proc. ICASSP* (Vol. 1, p. 788–791). Detroit, MI.

Mattera, D., Palmiera, F., and Haykin, S. (2000). Semiparametric support vector machines for nonlinear model estimation. In *Intelligent signal processing*. New York: IEEE Press.

McLachan, G. (1992). *Discriminant analysis and statistical pattern recognition*. New York: Wiley.

Nadaraya, E. A. (1964). On estimating regression. *Theor. Probab. Appl.*, *9*, 141–142.

Nadaraya, E. A. (1965). On nonparametric estimation of density functions and regression curves. *Theor. Probab. Appl.*, *10*, 186–190.

Ni, X.-F., and Simons, S. J. R. (1996). Off-line identification of nonlinear systems using structurally adaptive radial basis function networks. In *Proc. of the 35th IEEE Conference on Decision and Control* (Vol. 1, p. 935–936). Kobe, Japan: IEEE.

Niyogi, P., and Girosi, F. (1996). On the relationship between generalization error, hypothesis complexity, and sample complexity for radial basis functions. *Neural Computation*, *8*(4), 819–842.

Ormoneit, D. (1999). A regularization approach to continuous learning with an application to financial derivative pricing. *Neural Networks*, *12*, 1405–1412.

Oseledets, V. I. (1968). The multiplicative ergodic theory. Characteristic Lyapunov exponents for dynamical systems. *Trans. Moscow Math. Soc.*, *19*, 179–210.

O'Sullivan, F., Yandell, B., and Raynor, W. (1987). Automatic smoothing of regression functions in generalized linear models. *J. ASA*, *81*, 96–103.

Ott, E. (1993). *Chaos in dynamical systems*. Cambridge, UK: Cambridge University Press.

Ott, E., Sauer, T., and Yorke, J. A. (Eds.). (1994). *Coping with chaos: Analysis of chaotic data and the exploitation of chaotic systems*. New York: Wiley.

Parisini, T., and Zoppoli, R. (1994). Neural networks for nonlinear state estimation. *Intern. J. Robust and Nonlinear Control*, *4*, 231–248.

Parisini, T., and Zoppoli, R. (1996). Neural approximation for multistage optimal control of nonlinear stochastic systems. *IEEE Trans. Automatic Control*, *41*(6), 889–895.

Park, J., and Sandberg, I. W. (1991). Universal approximation using radial basis function networks. *Neural Computation*, *3*(2), 246–257.

Park, J., and Sandberg, I. W. (1993). Approximation and radial basis function networks. *Neural Computation*, *5*(2), 305–316.

Parzen, E. (1962). On estimation of a probability density function and mode. *Ann. Math. Statist.*, *33*, 1065–1076.

Perlmutter, B. A. (1989). Learning state trajectories in recurrent neural networks. *Neural Computation*, *1*, 263–269.

Pham, T. D., and Trans, L. T. (1991). Kernel density estimation under a locally mixing condition. In G. Roussas (Ed.), *Nonparametric functional estimation and related topics* (Vol. 335, p. 419–30). Dordrecht, Netherlands: Kluwer.

Pineda, F. J., and Sommerer, J. C. (1994). A fast algorithm for estimating the generalized dimension and choosing time delays. In A. S. Weigend and N. A. Gershenfeld (Eds.), *Time series prediction: Forecasting the future and understanding the past* (p. 367–385).

Platt, J. C. (1991). A resource allocating network for function interpolation. *Neural Computation*, *3*, 213–225.

Plutowski, M., Sakata, S., and White, H. (1994). Cross-validation estimates IMSE. In *Advances in neural information processing systems* (Vol. 6, p. 391–393). San Mateo, CA: Morgan Kaufman.

Poggio, T., and Girosi, F. (1990). Networks for approximation and learning. *Proc. IEEE*, *78*(9), 1484–1487.

Pollard, D. (1984). *Convergence of stochastic processes*. New York: Springer-Verlag.

Principe, J. C., and Kuo, J.-M. (1995). Dynamic modelling of chaotic time series with neural networks. In G. Tesauro, D. S., Touretzky, and T. Leen (Eds.), *Advances in neural information processing systems* (Vol. 7, p. 311–318). Cambridge, MA: MIT Press.

Principe, J. C., Rathie, A., and Kuo, J.-M. (1992). Prediction of chaotic time series with neural networks and the issue of dynamic modelling. *Intern. J. Bifurcation Chaos*, *2*(4), 989–996.

Rao, A. V., Miller, D., Ross, K., and Gersho, A. (1997). Mixture of experts regression modeling by deterministic annealing. *IEEE Trans. Signal Processing, 45*(11), 2811–2820.

Richard, M. D., and Lippmann, R. P. (1991). Neural network classifiers estimate Bayesian a posteriori probabilities. *Neural Computation, 3*, 461–483.

Rissanen, J. (1978). Modeling by shortest data description. *Automatica, 14*, 465–471.

Rosenblatt, M. (1956). Remarks on some nonparametric estimates of a density function. *Ann. Math. Statist., 27*, 832–837.

Ross, K. (1998). Deterministic annealing for clustering, compression, classification, regression, and related optimisation problems. *Proc. IEEE, 11*(86), 2210–2239.

Ross, K., Gurewitz, E., and Fox. G. C. (1992). Vector quantization by deterministic annealing. *IEEE Trans. Information Theory, 4*(38), 1249–1257.

Ruck, D., Rogers, S., Kabrisky, M., Oxley, M., and Suter, B. (1990). The multiplayer perceptron as an approximation to a Bayes optimal discriminant function. *IEEE Trans. Neural Networks, 1*(4), 296–298.

Rutkowski, L. (1985a). Nonparametric identification of quasi-stationary systems. *Systems and Controls Letters, 6*, 33–35.

Rutkowski, L. (1985b). Real-time identification of time-varying systems by non-parametric algorithms based on Parzen kernels. *Inter. J. Systems Science, 16*(9), 1123–1130.

Sauer, T., Yorke, J. A., and Casdagli, M. (1991). Embedology. *J. Statistical Physics, 65*(3/4), 579–616.

Scargle, J. D. (1992). Predictive deconvolution of chaotic and random processes. In D. Brillinger, P. Caines, J. Geweke, E. Parzen, M. Rosenblatt, and M. S. Taqu (Eds.), *New directions in time series analysis, part I* (Vol. 45, p. 335–356). New York: Springer-Verlag.

Schouten, J. C., and Bleek, C. M. van den. (1994). *RRCHAOS Time Series Analysis Software*. Delft University of Technology, Delft, Netherlands.

Schouten, J. C., Takens, F., and Bleek, C. M. van den. (1994). Estimation of the dimension of a noisy attractor. *Phys. Rev. E, 50*, 1851–1861.

Schroeder, M. R., and Atal, B. (1985). Code-excited linear prediction (CELP): high quality speech at very low bit rates. In *Proc. ICASSP* (p. 937). Tampa, FL.

Steck, J. E. (1992). Convergence of recurrent networks as contraction mappings. In *Proc. IJCNN* (Vol. 3, p. 462–467). Baltimore, MD.

Stinchombe, M., and White, H. (1990). Approximating and learning unknown mappings using multiplayer feedforward neural networks with bounded weights. In *Proc. IJCNN* (Vol. 3). Detroit, MI.

Stone, C. J. (1977). Consistent nonparametric regression. *Ann. Statist., 5*(4), 595–645.

Sun, M., and Glowinski, R. (1993). Pathwise approximation and simulation for the Zakai filtering equation through operator splitting. *Calcolo*, *30*, 219–239.

Takens, F. (1981). Detecting strange attractors in turbulence. In *Dynamical systems and turbulence* (Vol. 898, p. 366–381). New York: Springer-Verlag.

Terano, T., Asai, K., and Sugeno, M. (1992). *Fuzzy systems theory and its applications*. Boston: Academic Press.

Theiler, J., Eubank, S., Longtin, A., Galdrikian, B., and Farmer, J. D. (1992). Testing for nonlinearity in time series: the method of surrogate data. *Physica D*, *58*, 77–94.

Tong, H. (1990). *Non-linear time series: a dynamical systems approach*. Oxford Science.

Tong, H. (1992). Contrasting aspects of non-linear time analysis. In D. Brillinger, P. Caines, J. Geweke, E. Parzen, M. Rosenblatt, and M. S. Taqu (Eds.), *New directions in time series analysis, part I* (Vol. 45, p. 357–370). New York: Springer-Verlag.

Townshend, B. (1991). Nonlinear prediction of speech. In *Proc. ICASSP* (Vol. 1, p. 425–428). Toronto, Canada.

Van Ryzin, J. (1966). Bayes risk consistency of classification procedures using density estimates. *Sankhyā A*, *28*, 161–170.

Vapnik, V. (1982). *Estimation of dependences based on empirical data*. New York: Springer-Verlag.

Vapnik, V. (1992). Principles of risk minimization for learning theory. In D. Touretzky (Ed.), *Advances in neural information processing systems 4* (p. 831–838). San Mateo, CA: Morgan Kaufmann.

Villalobos, M., and Wahba, G. (1987). Inequality-constrained multivariate smoothing splines with application to the estimation of posterior probabilities. *J. ASA*, *82*(297), 239–248.

Wahba, G. (1990). *Spline models for observational data* (Vol. 59). SIAM.

Watson, G. S. (1964). Smooth regression analysis. *Sankhyā A*, *26*, 359–372.

White, H. (1989). Learning in artificial neural networks: a statistical perspective. *Neural Computation*, *1*, 425–464.

White, H. (1990). Connectionist nonparametric regression: multilayer feed-forward networks can learn arbitrary mappings. *Neural Networks*, *3*, 535–550.

Xu, L., Krzyżak, A., and Yuille, A. (1994). On radial basis function nets and kernel regression: statistical consistency, convergence rates, and receptive field size. *Neural Networks*, 7(4), 609–628.

Yee, P. (1992). *Classification experiments involving backpropagation and radial basis function networks* (CRL Report No. 249). McMaster University, Hamilton, ON: Communications Research Laboratory.

Yee, P., and Haykins, S. (1995). A dynamic, regularized Gaussian radial basis function network for nonlinear, nonstationary time series prediction. In *Proc. ICASSP* (p. 3419–3422). Detroit, MI.

Zemanian, A. H. (1987). *Distribution theory and transform analysis: an introduction to generalized functions, with applications*. Dover.

Zheng, N., Zhang, Z., Zheng, H., and Gang, S. (2000). Deterministic annealing learning of the radial basis function nets for improving the regression ability of RBF network. In *Proc. IEE-INNS-ENNS IJCNN.* (vol. 3, p. 601–607) Como, Italy .

Ziemer, W. (1989). *Weakly differentiable functioins* (Vol. 120). New York: Springer-Verlag.

ADDITIONAL BIBLIOGRAPHY

Arnott, R. (1997). Diversity combining for digital mobile radio using radial basis function networks. *Signal Processing*, *63*, 1–16.

Bianchini, M., Frasconi, P., and Gori, M. (1995). Learning without local minima in radial basis function networks. *IEEE Trans. Neural Networks*, *6*, 749–756.

Blanzieri, E., and Katenkamp, P. (1996). Learning radial basis function networks on-line. In *Machine learning, proc. of the 13th international conference* (p. 37–45). Bari, Italy: Morgan Kaufman.

Cancelo, G., and Mayosky, M. (1998). A parallel analog signal processing unit based on radial basis function networks. *IEEE Trans. Nuclear Science*, *45*, 792–797.

Canete, F. de, Garcia-Cerezo, A., and Garcia-Moral, I. (1998). Direct control with radial basis function networks: stability analysis and applications. *J. Systems Architecture*, *44*, 583–596.

Ceccarelli, M., and Hounsou, J. T. (1996). Sequence recognition with radial basis function networks: experiments with spoken digits. *Neurocomputing*, *11*, 75–88.

Frasconi, P., Gori, M., Maggini, M., and Soda, G. (1996). Representation of finite state automata in recurrent radial basis function networks. *Machine Learning*, *23*, 5–32.

Freeman, A., and Saad, D. (1995). Learning and generalization in radial basis function networks. *Neural Computation*, *7*, 1000–1020.

Freeman, A., and Saad, D. (1997). Online learning in radial basis function networks. *Neural Computation*, *9*, 1601–1622.

Freeman, J., and Saad, D. (1997). Dynamics of on-line learning in radial basis function networks. *Phys. Rev. E (Statistical Physics, Plasma, Fluids, and Related Interdisciplinary Topics)* *56*, 907–918.

Junge, F., and Unbehauen, H. (1996). Off-line identificiation of nonlinear systems using structurally adaptive radial basis function networks. In *Proc. of the 35th IEEE Conference on Decision and Control* (Vol. 1, p. 943–948). Kobe, Japan: IEEE.

Kiernan, L., Mason, J. D., and Warwick, K. (1996). Robust initialisation of Gaussian radial basis function networks using partitioned *k*-means clustering. *Electronics Letters*, *32*, 671–673.

Krzyżak, A., and Linder, T. (1996a). Radial basis function networks and complexity regularization in function learning. In *Advances in neural information processing systems* (Vol. 9, p. 197–203). Denver, CO: MIT Press.

Krzyżak, A., and Linder, T. (1996b). Radial basis function networks and nonparametric classification: complexity regularization and rates of convergence. In *Proc. of the 13th International Conference on Pattern Recognition* (Vol. 4, p. 650–653). Vienna, Austria: IEEE Comput. Soc. Press.

Krzyżak, A., and Linder, T. (1998). Radial basis function networks and complexity regularization in function learning. *IEEE Trans. Neural Networks*, *9*, 247–256.

Langari, R., Liang, W., and Yen, J. (1997). Radial basis function networks, regression weights, and the expectation-maximization algorithm. *IEEE Trans. Systems, Man & Cybernetics, Part A (Systems & Humans)*, *27*, 613–623.

Liu, P., Kadirkamanathan, V., and Billings, S. A. (1996). Stable sequential identification of continuous nonlinear dynamical systems by growing radial basis function networks. *Intern. J. Control*, *65*, 53–69.

Mallofre, C., and Llanas, X. P. (1996). Fault tolerance parameter model of radial basis function networks. In *Proc. ICNN* (Vol. 2, p. 1384–1389). Washington, DC: IEEE.

Mazurek, J., Krzyżak, A., Cichocki, A. (1997). Radial basis function networks and complexity regularization in function learning. In *Proc. ICASSP* (Vol. 4, p. 3317–20). Munich, Germany: IEEE Comput. Soc. Press.

Nelles, O., and Isermann, R. (1995). Identification of nonlinear dynamic systems: classical methods versus radial basis function networks. In *Proc. of the 1995 American Control Conference* (Vol. 5, p. 3786–3790). Seattle, WA: American Autom. Control Council.

Ogawa, S., Ikeguchi, T., Matozaki, T., and Aihra, K. (1996). Nonlinear modelling by radial basis function networks. *IEICE Trans. Fund. Electronics, Communications and Computer Sciences*, *E79-A*, 1608–1617.

Salomon, R. (1997). Radial basis function networks for autonomous agent control. In *Proc. ICNN* (Vol. 3, p. 1868–1872). Houston, TX: IEEE.

Sanner, M., and Essex, C. (1996). Multiresolution radial basis function networks for the adaptive control of robotic systems. In *UKACC international conference on control* (Vol. 2, p. 894–898). Exeter, UK: IEE.

Schram, G., Tempel, M. W. V. D., and Klingsman, A. J. (1995). Structure determination of radial basis function networks with an application to flight control. In *Artificial Intelligence in Real-Time Control 1995* (p. 95–100). Bled, Slovenia: Pergamon.

Tokhi, O., and Wood, R. (1997). Active noise control using radial basis function networks. *Control Engineering Practice*, *5*, 1311–1322.

Zaid, F. Ahmed, Ioannou, P. A. Polycarpou, M. M., and Yousef, M. M. (1993). Identification and control of aircraft dynamics using radial basis function networks. In *Proc. 2nd IEEE Conf. Control Applications* (Vol. 2, p. 567–572). Vancouver, BC: IEEE.

Zooghby, H. El, Christodoulou, C. G., and Georgiopoulos, M. (1997). Performance of radial-basis function networks for direction of arrival estimation with antenna arrays. *IEEE Trans. Antennas and Propagation*, *45*, 1611–1617.

INDEX

9 780471 353492